W9-CFV-877

# Study Guide For
## Third Edition Of
# ESSENTIALS
## OF FIRE FIGHTING

**Developed by**
**Marsha Sneed**
**Publications Specialist**

**Published by**
**Fire Protection Publications**
**Oklahoma State University**

*ISBN 0-87939-102-2*
*Library of Congress 92-073546*

*First Edition*
*First Printing, December 1992*
*Second Printing, April 1993*

*Printed in the United States of America*

# Dedication

*This manual is dedicated to the members of that unselfish organization
of men and women who hold devotion to duty
above personal risk, who count on sincerity of service above
personal comfort and convenience, who strive unceasingly to find
better ways of protecting the lives, homes and property
of their fellow citizens from the ravages of fire and other
disasters . . .* **The Firefighters of All Nations.**

Dear Firefighter:

The International Fire Service Training Association (IFSTA) is an organization that exists for the purpose of serving firefighters' training needs. Fire Protection Publications is the publisher of IFSTA materials. Fire Protection Publications staff members participate in the National Fire Protection Association, International Society of Fire Service Instructors, and the International Association of Fire Chiefs.

If you need additional information concerning our organization or assistance with manual orders, contact:

> **Customer Services**
> **Fire Protection Publications**
> **Oklahoma State University**
> **Stillwater, OK 74078-0118**
> **1 (800) 654-4055**

For assistance with training materials, recommended material for inclusion in a manual, or questions on manual content, contact:

> **Technical Services**
> **Fire Protection Publications**
> **Oklahoma State University**
> **Stillwater, OK 74078-0118**
> **(405) 744-5723**

# NOTICE

The questions in this study guide are taken from the information in the third edition of **Essentials Of Fire Fighting**, an IFSTA-validated manual. The questions are *not validated test questions and are not intended to be duplicated or used for certification or promotional examinations*; this guide is intended to be used as a tool for studying the information presented in **Essentials Of Fire Fighting**.

# Table Of Contents

# Preface

This study guide is designed to help the reader understand the material presented in the IFSTA-validated **Essentials Of Fire Fighting**, third edition. It identifies important information and concepts from each chapter and provides questions to help reinforce them.

When used properly, this study guide ensures a better understanding of the knowledge and of those skills and tasks required by NFPA 1001, *Standard for Fire Fighter Professional Qualifications*, Levels I and II. This study guide serves as an excellent study reference for firefighters who are preparing for certification or promotional examinations.

Thank-you to the following members of the Fire Protection Publications staff whose contributions made the publication of this study guide possible.

Marsha Sneed, Publications Specialist
Michael A. Wieder, Senior Publications Editor
Susan S. Walker, Coordinator of Instructional Development
Barbara Adams, Publications Specialist
Cynthia Brakhage, Publications Specialist
Don Davis, Production Coordinator
Ann Moffat, Senior Graphic Designer
Desa Porter, Graphic Designer
Lori Schoonover, Graphic Designer
Karen Murphy, Phototypesetting Technician

Lynne C. Murnane
Managing Editor

# How To Use This Book

This study guide is designed to be used in conjunction with and as a supplement to the third edition of the IFSTA-validated **Essentials Of Fire Fighting** manual. The questions are designed to make you think — they are *not* designed to trick or mislead you. To receive the maximum learning experience from this study guide, it is recommended that you use the following procedure:

1. Read one chapter at a time in the **Essentials Of Fire Fighting** manual. After reading a chapter, underline or highlight important terms, topics, and subject matter in that chapter.

2. Open the study guide to the corresponding chapter. Answer all the questions in the study guide for that chapter. After you have completed, check your answers with those in the answer section at the end of the study guide.

   **NOTE:** *DO NOT* answer each question and then immediately check the answer for the correct response.

   If you find that you have answered any question incorrectly, find the explanation of the answer in the **Essentials Of Fire Fighting** manual. The number in parentheses after each answer in the answer section identifies the page where the answer can be found. Correct any incorrect answers, and review the material that was answered incorrectly.

3. Go to the next chapter of the manual, and repeat Steps 1 and 2.

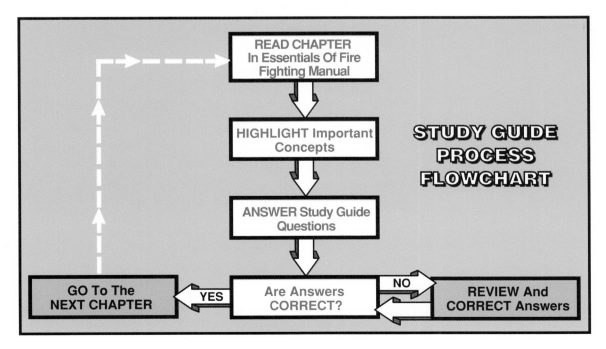

1

# Fire Behavior

# Fire Behavior | 1

## DEFINITIONS OF KEY TERMS

**Define each of the following terms.**

1. Combustion

   _____
   _____

2. Fire

   _____
   _____

3. British thermal unit (Btu)

   _____
   _____

4. Calorie

   _____
   _____

5. Celsius (Centigrade)

   _____
   _____

6. Fahrenheit

   _____
   _____

7. Fire point

   _____
   _____

8. Flame spread

   _____
   _____

**1**

9. Flash point

_____

_____

10. Heat

_____

_____

11. Ignition temperature

_____

_____

12. Oxidation

_____

_____

13. Conduction

_____

_____

14. Convection

_____

_____

15. Radiation

_____

_____

16. Pyrolysis

_____

_____

17. Specific gravity

_____

_____

18. Vapor density

_____

_____

## TRUE/FALSE                                                                    **1**

**Mark each of the following statements true (T) or false (F). Correct each false statement.**

19. ☐ T   ☐ F   Mechanical heat is generated two ways — by friction and by compression.

_____

_____

20. ☐ T   ☐ F   Nuclear heat energy is generated when atoms are either split apart or combined.

_____

_____

21. ☐ T   ☐ F   Of the three states of matter in which fuel may be found, only gases and solids burn.

_____

_____

22. ☐ T   ☐ F   Radiated heat is one of the major sources of fire spread to exposures.

_____

_____

23. ☐ T   ☐ F   If a solid fuel is in a horizontal position, fire spread will be more rapid than if it is in a vertical position.

_____

_____

24. ☐ T   ☐ F   If heat is dissipated faster than it is generated, a positive heat balance is created; a positive heat balance is required to maintain combustion.

_____

_____

25. ☐ T   ☐ F   Flame is the product of combustion that is responsible for the spread of fire.

_____

_____

26. ☐ T   ☐ F   If a fire is in the smoldering mode of combustion, only three extinguishment options exist: reduction of temperature, elimination of fuel, or elimination of oxygen.

_____

_____

**1**

27. ☐ T  ☐ F   Fires involving low flash point liquids and flammable gases cannot be extinguished by cooling with water.

_____

_____

## MATCHING

**Match the following types of electrical heat with their correct descriptions. Write the correct letters in the blanks.**

28. _____ Heating that occurs as a result of the action of pulsating either direct current (DC) or alternating current (AC) at high frequency on a nonconductive material

29. _____ Heat resulting from the buildup of a positive charge on one surface and a negative charge on another surface

30. _____ Heating that occurs when the current flow is interrupted

31. _____ Heat generated by passing an electrical current through a conductor such as a wire or an appliance

32. _____ Heating that occurs when a wire is not insulated well enough to contain all the current

A. Resistance heating
B. Dielectric heating
C. Leakage current heating
D. Heat from arcing
E. Static electricity

## IDENTIFICATION

**Identify the following items.**

33. Boiling point

_____

_____

34. Flammable or explosive limits

_____

_____

35. Vapor pressure

_____

_____

**1**

36. The Law of Heat Flow

_____

_____

37. Radiative feedback

_____

_____

**Provide the requested information.**

38. Identify the types of chemical heat energy, and give a brief description of each.

A. _____

_____

B. _____

_____

C. _____

_____

D. _____

_____

39. Describe the three phases of fire.

A. Incipient phase

_____

_____

_____

_____

_____

_____

B. Steady-state burning phase

_____

_____

_____

_____

_____

_____

C. Hot-smoldering phase

_____

_____

_____

_____

_____

_____

**1**

## LISTING

40. List at least five conditions that may indicate the potential for a backdraft.

_____

_____

_____

_____

_____

41. List the four products of combustion.
    A. _____
    B. _____
    C. _____
    D. _____

42. List fuels involved in Class A fires.

_____

_____

43. List fuels involved in Class B fires.

_____

_____

44. List fuels involved in Class C fires.

_____

_____

45. List fuels involved in Class D fires.

_____

_____

## SHORT ANSWER

**Briefly answer each question in your own words.**

46. With the exception of methane, every hydrocarbon has a vapor density greater than one. What does this tell the firefighter?

_____

_____

_____

_____

47. What do the four sides of the fire tetrahedron represent?

_____

_____

_____

_____

48. What are the differences and similarities of the fire triangle and the fire tetrahedron?

_____

_____

_____

_____

49. Under normal conditions, what percent of the air is oxygen? What does an oxygen concentration below this percentage mean to the firefighter?

_____

_____

_____

_____

50. What is the difference between rollover and flashover?

_____

_____

_____

_____

_____

51. What is thermal layering, and why is it important that thermal layering not be disrupted?

_____

_____

_____

_____

52. Describe the extinguishment method for Class A fires.

_____

_____

_____

_____

**1**

53. Describe the extinguishment method for Class B fires.

_____

_____

_____

_____

_____

54. Describe the extinguishment method for Class C fires.

_____

_____

_____

_____

_____

55. Describe the extinguishment method for Class D fires.

_____

_____

_____

_____

_____

**2**

# Portable
# Extinguishers

# Portable Extinguishers | 2

**Mark each of the following statements true (T) or false (F). Correct each false statement.**

1. ☐ T ☐ F Portable fire extinguishers for Class A, Class B, and Class C fires have a letter rating and a numerical rating; those for Class D fires have only a letter rating.

   _____

   _____

2. ☐ T ☐ F The effectiveness of an extinguisher rated for a Class D fire is indicated on a paper tag attached to the extinguisher.

   _____

   _____

3. ☐ T ☐ F For a Class C designation of a portable extinguisher, the extinguishing agent is only tested for electrical nonconductivity.

   _____

   _____

4. ☐ T ☐ F Class D agents cannot be given a multipurpose rating to be used on other classes of fire.

   _____

   _____

5. ☐ T ☐ F Modern fire extinguishers are designed to be carried to the fire in an upright position and then turned upside down before operation.

   _____

   _____

6. ☐ T ☐ F Dry chemical extinguishers are among the most common portable fire extinguishers in use today.

   _____

   _____

**2**

7. ☐ T ☐ F   Leaking, corroded, or otherwise damaged extinguisher shells or cylinders should be repaired by local fire department personnel before being placed back into service.

_____

_____

8. ☐ T ☐ F   Servicing portable fire extinguishers is the responsibility of the local fire department.

_____

_____

9. ☐ T ☐ F   Fire extinguishers should be thoroughly inspected at least once a year.

_____

_____

## MULTIPLE CHOICE

**Circle the letter before the most appropriate response.**

10. Portable extinguishers to be used on Class A and Class B fires have a numerical rating that indicates _____.
    A. The number of fires the extinguisher can effectively extinguish before it is recharged
    B. The extinguishing potential or relative effectiveness of the extinguisher
    C. The number of times per year that the extinguisher must be inspected
    D. The type of fire that can be extinguished with the extinguisher

11. Class A portable fire extinguishers are rated _____, depending on their size.
    A. From 1-A through 10-A
    B. From 1-A through 6-A
    C. From 1-A through 40-A
    D. From 1-A through 100-A

12. In order for a Class A portable fire extinguisher to be rated 1-A, _____ of water is/are required. To be rated 4-A, _____ of water are required.
    A. 1¼ gallons (5 L); 5 gallons (20 L)
    B. ½ gallon (2 L); 2 gallons (8 L)
    C. 1 gallon (4 L); 2 gallons (8 L)
    D. 1 gallon (4 L); 4 gallons (16 L)

**2**

13. Extinguishers suitable for use on Class B fires have numerical ratings ranging _____.
    A. From 1-B through 10-B
    B. From 6-B through 460-B
    C. From 1-B through 640-B
    D. From 4-B through 6-B

14. If an extinguisher shows only slight damage or corrosion and it is questionable whether it is safe to use, it should be _____.
    A. Discarded
    B. Given a hydrostatic test by the manufacturer or a qualified testing agency
    C. Inspected and tested by specially trained fire department personnel
    D. Placed back into service if it can be recharged successfully

## MATCHING

**Match the following fire extinguisher geometric symbols to their applicable types of fire. Write the correct letters in the blanks.**

15. _____ Ordinary combustible materials fire

16. _____ Combustible metal fire

17. _____ Electrical equipment fire

18. _____ Flammable liquid fire

A.

B.

C.

D.

**2**

**Match the following fire extinguisher symbols to their correct interpretations. Write the correct letters in the blanks.**

19. _____ Extinguisher suitable for Class B and Class C fires but not for Class A fires

20. _____ Extinguisher suitable for Class A and Class B fires but not for Class C fires

21. _____ Extinguisher suitable for Class A, Class B, and Class C fires

22. _____ Extinguisher suitable for Class A fires

A.

B.

C.

D.

## IDENTIFICATION

**Identify the following items.**

23. NFPA 10

_____

24. Underwriter's Laboratories, Inc. (UL) and Underwriter's Laboratories of Canada (ULC)

_____

25. AFFF

_____

26. Halon 1211

_____

**2**

27. Halon 1301

_____

28. Dry chemical agents

_____

29. Dry powder agents

_____

**Provide the requested information.**

30. The general operating instructions for operating portable fire extinguishers follow the letters P-A-S-S. Identify the steps.

P — _____

_____

A — _____

_____

S — _____

_____

S — _____

_____

**Provide the requested information for the following portable fire extinguishers when operating under normal conditions.**

31. Pump tank water extinguishers
    A. Class of fire suited for _____
    B. Size _____
    C. Stream reach _____
    D. Discharge time _____
    E. Basic operation_____

    _____

    _____

    _____

    _____

32. Stored-pressure water extinguishers
    A. Class of fire suited for _____
    B. Size _____
    C. Stream reach _____
    D. Discharge time _____
    E. Basic operation_____

    _____

    _____

**2**

33. Aqueous film forming foam (AFFF) extinguishers
    A. Class of fire suited for _____
    B. Size _____
    C. Stream reach _____
    D. Discharge time _____
    E. Basic operation _____

    _____
    _____
    _____
    _____

34. Halon 1211 extinguishers
    A. Class of fire suited for _____
    B. Size _____
    C. Stream reach _____
    D. Discharge time _____
    E. Basic operation _____

    _____
    _____
    _____
    _____

35. Halon 1301 extinguishers
    A. Class of fire suited for _____
    B. Size _____
    C. Stream reach _____
    D. Basic operation _____

    _____
    _____
    _____
    _____

**2**

36. Carbon dioxide extinguishers (hand carried)
    A. Class of fire suited for _____
    B. Size _____
    C. Stream reach _____
    D. Discharge time _____
    E. Basic operation _____

    _____

    _____

    _____

    _____

37. Carbon dioxide wheeled units
    A. Class of fire suited for _____
    B. Size _____
    C. Stream reach _____
    D. Discharge time _____
    E. Basic operation _____

    _____

    _____

    _____

    _____

38. Dry chemical extinguishers (hand carried)
    A. Class of fire suited for _____
    B. Size _____
    C. Stream reach _____
    D. Discharge time _____
    E. Basic operation _____

    _____

    _____

    _____

    _____

    _____

**2**

39. Dry chemical wheeled units
    A. Class of fire suited for _____
    B. Size _____
    C. Stream reach _____
    D. Discharge time _____
    E. Basic operation _____

    _____

    _____

    _____

    _____

## LISTING

40. List the three Class A combustibles that an extinguisher must be capable of extinguishing in order to receive a 1-A through a 6-A rating.
    A. _____
    B. _____
    C. _____

41. Test fires for establishing Class D ratings vary with the type of combustible metal being tested. List the factors considered during each test.

    _____

    _____

    _____

    _____

42. List the factors to consider when selecting the proper portable fire extinguisher.

    _____

    _____

    _____

    _____

43. Fire extinguishers must be inspected regularly to ensure that they are accessible and operable. This is done by verifying what three things?
    A. _____

    _____

    B. _____

    _____

    C. _____

    _____

2

44. All maintenance procedures should include a thorough examination of the three basic parts of an extinguisher. List these.

A. _____

B. _____

C. _____

## SHORT ANSWER

**Briefly answer each question in your own words.**

45. How are portable fire extinguishers classified?

_____

_____

46. On what is the rating of a Class B extinguisher based?

_____

_____

47. Describe the two methods of labeling portable fire extinguishers.

_____

_____

_____

_____

_____

_____

48. NFPA 10 requires that pump tank water extinguishers, stored-pressure water extinguishers, and aqueous film forming foam extinguishers be protected against freezing if they are going to be exposed to temperatures less than 40°F (4°C). How is this accomplished?

_____

_____

49. Describe the application of an extinguishing agent that is used to extinquish burning metal.

_____

_____

_____

_____

**2**

50. Describe the application of dry powder agents to a metal fire when the burning metal is on a combustible surface.

_____

_____

_____

_____

51. Why is it important that firefighters be able to recognize obsolete fire extinguishers?

_____

_____

_____

_____

3

# Firefighter Personal Protective Equipment

# Firefighter Personal Protective Equipment | 3

## DEFINITIONS OF KEY TERMS

**Define each of the following terms.**

1. Pulmonary edema

   _____

   _____


2. Synergistic effect

   _____

   _____


3. Open-circuit SCBA

   _____

   _____


4. Closed-circuit SCBA

   _____

   _____


5. Cascade system

   _____

   _____


6. Skip breathing

   _____

   _____


## TRUE/FALSE

**Mark each of the following statements true (T) or false (F). Correct each false statement.**

7. ☐ T  ☐ F   Helmets must have earflaps, which should always be used during fire fighting.

   _____

   _____

**3**

8. ☐ T ☐ F   Faceshields attached to a helmet must be approved by the manufacturer of the helmet.

_____

_____

9. ☐ T ☐ F   Work uniforms that meet NFPA 1975 are designed to be fire resistant and are designed to be worn as protective clothing for fire fighting operations.

_____

_____

10. ☐ T ☐ F   When oxygen concentrations are below 18 percent, the human body responds by decreasing the respiratory rate.

_____

_____

11. ☐ T ☐ F   The tissue damage resulting from inhaling hot air is immediately reversible by introducing fresh, cool air.

_____

_____

12. ☐ T ☐ F   More fire deaths occur from exposures to carbon monoxide than from exposures to any other toxic product of combustion.

_____

_____

13. ☐ T ☐ F   The physical factors that affect the firefighter's ability to use SCBA effectively are physical condition, agility, and facial features.

_____

_____

14. ☐ T ☐ F   Proper respiratory functioning will maximize the wearer's operation time in a self-contained breathing apparatus.

_____

_____

15. ☐ T ☐ F   Steel cylinders are the most common cylinders and are 30 percent stronger than other types.

_____

_____

**3**

16. ☐ T ☐ F  The SCBA regulator pressure gauge should read within 10 psi (70 kPa) of the SCBA cylinder gauge if increments are in psi (kPa).

_____

_____

17. ☐ T ☐ F  On SCBA with nonfacepiece-mounted regulators, air from the cylinder travels to the regulator through the high-pressure hose and travels from the regulator to the facepiece through the low-pressure hose.

_____

_____

18. ☐ T ☐ F  When preparing to don the SCBA backpack using the crossed-arms coat method, the firefighter should crouch or kneel at the end of the cylinder opposite the cylinder valve.

_____

_____

19. ☐ T ☐ F  When using the regular coat method for donning the SCBA backpack (with the regulator on the left side), the firefighter should grasp the left shoulder strap high on the harness with the left hand and should grasp the lower portion of the same strap (or grasp the regulator) with the right hand.

_____

_____

20. ☐ T ☐ F  Donning SCBA en route is accomplished by inserting the arms through the straps while sitting with the seat belt on and then adjusting the straps for a snug fit.

_____

_____

21. ☐ T ☐ F  The side- or rear-mounted SCBA allows several steps of the donning procedure to be eliminated and permits donning en route.

_____

_____

22. ☐ T ☐ F  Facepiece harness straps should be tightened by pulling them evenly and simultaneously to the rear.

_____

_____

23. ☐ T ☐ F  All firefighters who wear SCBA should be certified as physically fit by a physician using criteria established by the fire department.

_____

_____

**3**

**Circle the letter before the most appropriate response.**

24. Some of the best materials available for eye protection are made of _____.
    A. Elastomer
    B. Polyvinyl chloride
    C. Acrylic-coated shatterproof glass
    D. Polycarbonates

25. Firefighters who wear prescription safety eyeglasses should select frames and lenses that meet _____.
    A. NFPA 1500
    B. NFPA 1981
    C. ANSI Z87.1
    D. ANSI Z88.5

26. Firefighters A and B are discussing reflective trim on turnout gear. Firefighter A says that to meet NFPA standards, each coat must have at least 325 square inches (209 700 mm²) of reflective tape. Firefighter B agrees but says that protective trousers are not required to have reflective trim. Who is correct?
    A. Firefighter A
    B. Firefighter B
    C. Both A and B
    D. Neither A nor B

27. Firefighters A and B are discussing plastic-coated gloves. Firefighter A says that they are useful during overhaul and hose pickup operations. Firefighter B says that such gloves are *not* recommended for fire fighting. Who is correct?
    A. Firefighter A
    B. Firefighter B
    C. Both A and B
    D. Neither A nor B

28. Which of the following statements about protective coats and trousers is correct?
    A. They can all be cleaned with special dry-cleaning solvent.
    B. They cannot be cleaned by hand-scrubbing.
    C. They should be cleaned according to the manufacturer's directions.
    D. None of the above statements are correct.

29. OSHA requires workers to wear protective breathing equipment whenever entering an atmosphere that has an oxygen concentration of less than _____.
    A. 19½ percent
    B. 21 percent
    C. 25 percent
    D. 87½ percent

30. Firefighter A says that SCBA should be worn when performing rescues in sewers and manholes. Firefighter B says that SCBA should be worn when entering silos, storage tanks, and tank cars. Who is correct?
    A. Firefighter A
    B. Firefighter B
    C. Both A and B
    D. Neither A nor B

31. Firefighter A says that the interchanging of SCBA components with those of another manufacturer is NIOSH approved, providing such is outlined in the department's SOPs. Firefighter B says that NIOSH certification of SCBA is *not* voided by interchanging only cylinders and backpacks of different manufacturers. Who is correct?
    A. Firefighter A
    B. Firefighter B
    C. Both A and B
    D. Neither A nor B

32. Fully charged, the typical 30-minute-rated cylinder contains _____ of breathing air at 2,216 psi (15 290 kPa).
    A. 35 cubic feet (990 L)
    B. 45 cubic feet (1 275 L)
    C. 30 cubic feet (850 L)
    D. 40 cubic feet (1 133 L)

33. Cylinders rated for 60 minutes contain _____ of breathing air at 4,500 psi (31 000 kPa) when fully charged.
    A. 70 cubic feet (1 982 L)
    B. 90 cubic feet (2 550 L)
    C. 88 cubic feet (2 490 L)
    D. 67 cubic feet (1 900 L)

34. Which of the following is a function of the SCBA regulator?
    A. To reduce the pressure of the cylinder air to slightly above atmospheric pressure
    B. To keep the diaphragm closed so that the positive pressure of the facepiece assembly is not disturbed
    C. To control the flow of air to meet the respiratory requirements of the wearer
    D. Both A and C

35. During normal operation of SCBA, the mainline valve is _____, and the bypass valve is _____.
    A. Fully open and locked; closed
    B. Fully open and locked; fully open
    C. Closed; fully open and locked
    D. Closed; partially open

**3**

36. Airline equipment enables the firefighter to travel up to _____ from the regulated air supply source.
   A. 200 feet (60 m)
   B. 300 feet (90 m)
   C. 400 feet (120 m)
   D. 500 feet (150 m)

37. What should the firefighter do when checking the SCBA unit before donning if the audible alarm does not sound or if it sounds but does not stop?
   A. Fill the cylinder.
   B. Crack the bypass valve.
   C. Tag the unit and place it out of service.
   D. Open the cylinder valve.

38. When using the over-the-head method to don an SCBA, which of the following should you do first?
   A. Grasp the backplate or cylinder with both hands, one at each side.
   B. Adjust the shoulder straps to ensure that the SCBA harness will fit comfortably.
   C. Check the unit, including the cylinder gauge.
   D. Fasten the regulator to the harness if the regulator is harness-mounted.

39. Firefighters A and B are discussing checking the facepiece seal after donning the facepiece. Firefighter A says that the end of the low-pressure hose must be covered before inhaling. Firefighter B says that if the SCBA has a facepiece-mounted regulator, the opening on the facepiece for the regulator should be covered with one hand before inhaling. Who is correct?
   A. Firefighter A
   B. Firefighter B
   C. Both A and B
   D. Neither A nor B

40. Firefighters A and B are discussing confined-space entry. Firefighter A says that airline breathing equipment is very well suited for use in confined spaces. Firefighter B say that SCBA can be used, but the firefighter must remove the backpack, enter the confined space, and then redon the backpack. Who is correct?
   A. Firefighter A
   B. Firefighter B
   C. Both A and B
   D. Neither A nor B

## IDENTIFICATION

**Identify the following items.**
41. NFPA 1972

42. NFPA 1500

_____

43. NFPA 1971

_____

44. NFPA 1973

_____

45. NFPA 1974

_____

46. NFPA 1975

_____

47. CO

_____

48. HCl

_____

49. HCN

_____

50. $CO_2$

_____

51. $NO_2$

_____

52. $COCl_2$

_____

**Identify the functions of the following personal protective equipment.**

53. Helmet_____

_____

54. Protective hood _____

_____

55. Protective coat and trousers _____

_____

**3**

56. Gloves _____
_____

57. Safety shoes or boots _____
_____

58. Eye protection (goggles or faceshields) _____
_____

59. Hearing protection _____
_____

60. Self-contained breathing apparatus (SCBA) _____
_____

61. Personal Alert Safety System (PASS) _____
_____

**Provide the requested information for each of the following fire gases.**

62. Carbon monoxide
  A. Source _____
  _____

  B. Characteristics _____
  _____

  C. Effects on the body_____
  _____

63. Hydrogen chloride
  A. Source _____
  _____

  B. Characteristics _____
  _____

  C. Effects on the body_____
  _____

64. Hydrogen cyanide
  A. Source _____
  _____

  B. Characteristics _____
  _____

  C. Effects on the body_____
  _____

**3**

65. Carbon dioxide

   A. Source _____

   _____

   B. Characteristics _____

   _____

   C. Effects on the body _____

   _____

66. Nitrogen dioxide

   A. Source _____

   _____

   B. Characteristics _____

   _____

   C. Effects on the body _____

   _____

67. Phosgene

   A. Source _____

   _____

   B. Characteristics _____

   _____

   C. Effects on the body _____

   _____

**Identify how SCBA equipment limits the firefighter.**

68. Limited visibility _____

   _____

69. Decreased ability to communicate _____

   _____

70. Increased weight _____

   _____

71. Decreased mobility _____

   _____

**Identify the limitations of the SCBA air supply.**

72. Physical condition of user _____

   _____

**3**

73. Degree of physical exertion _____

_____

74. Emotional stability _____

_____

75. Condition of apparatus _____

_____

76. Cylinder pressure before use _____

_____

77. Training and experience _____

_____

**Provide the requested information.**

78. Firefighters need two different kinds of foot protection. What are the two types, and for what activities is each type appropriate?

    A. _____

       _____

    B. _____

       _____

## LISTING

79. Coats that meet NFPA 1971 have labels attached to each separable layer. What three items of information do these labels contain?

    A. _____

    B. _____

    C. _____

80. NFPA requires that all turnout coats be made of three layers. List these.

    A. _____

    B. _____

    C. _____

81. List at least five guidelines for maintaining helmets.

_____

_____

_____

_____

_____

_____

_____

_____

_____

_____

82. List the four guidelines for the care and maintenance of fire fighting boots.

A. _____

_____

B. _____

_____

C. _____

_____

D. _____

_____

83. List the four common hazardous atmospheres associated with fires or other emergencies.

A. _____

B. _____

C. _____

D. _____

84. The particular toxic gases given off at a fire vary according to four factors. List these factors.

A. _____

B. _____

C. _____

D. _____

**3**

85. List the four basic SCBA component assemblies.

A. _____

B. _____

C. _____

D. _____

86. List the three methods used to prevent or control internal fogging of an SCBA facepiece lens.

A. _____

B. _____

C. _____

87. List safety checks that should be made when preparing to don an SCBA.

_____

_____

_____

_____

_____

_____

_____

_____

_____

88. List checks that should be made during daily inspections of SCBA.

_____

_____

_____

_____

_____

_____

_____

_____

_____

_____

_____

_____

_____

_____

**3**

89. List the different methods that a firefighter can use to find a way out during an emergency.

_____

_____

_____

_____

_____

_____

_____

_____

_____

_____

## SHORT ANSWER

**Briefly answer each question in your own words.**

90. What are the potential hazards associated with wearing earplugs and earmuffs in structural fire fighting situations?

_____

_____

91. Why is it important to clean outer shells of protective clothing regularly?

_____

_____

92. What can result if excessively heated air (temperatures exceeding 120°F to 130°F [49°C to 54°C]) is inhaled?

_____

_____

_____

_____

_____

93. What is smoke?

_____

_____

**3**

94. What is the advantage of positive-pressure breathing apparatus over the demand type?

_____

_____

95. All modern SCBA units have an audible alarm that sounds when the cylinder pressure decreases to a preset level. When it sounds, what does the alarm indicate to the firefighter?

_____

_____

_____

_____

96. How does the SCBA regulator assembly work?

_____

_____

_____

_____

_____

_____

97. What are two important differences in the various types of facepiece assemblies?

_____

_____

_____

_____

98. How should SCBA facepiece assemblies be cleaned?

_____

_____

_____

_____

99. How often should SCBA equipment be removed from service in order to check the pressure regulator, valves, gauges, harness, and facepiece?

_____

100. How often should SCBA air cylinders be hydrostatically tested?

_____

_____

**3**

101. What safety precautions should a firefighter observe when filling an SCBA cylinder?

_____

_____

_____

_____

_____

102. In what instances is the bypass valve on the SCBA regulator used?

_____

_____

_____

_____

103. Explain the operation of the PASS device.

_____

_____

_____

_____

# 4

# Ropes
# And
# Knots

# Ropes And Knots | 4

## DEFINITIONS OF KEY TERMS

**Define each of the following terms.**

1. Life safety rope

_____

_____

2. Utility rope

_____

_____

## TRUE/FALSE

**Mark each of the following statements true (T) or false (F). Correct each false statement.**

3. ☐ T ☐ F  The two basic types of materials used to construct fire service rope are natural fibers and synthetic fibers.

_____

_____

4. ☐ T ☐ F  The tensile strength of Type #1 manila rope is comparable to that of nylon.

_____

_____

5. ☐ T ☐ F  According to NFPA 1983, life safety rope may be used only once and then taken out of service as a life safety rope.

_____

_____

6. ☐ T ☐ F  Knots and hitches for fire service use should be those that may be rapidly tied, can be easily untied, are not subject to slippage, and have a minimum of abrupt bends.

_____

_____

**4**

7. ☐ T ☐ F   The figure of eight on a bight knot is not a secure knot on synthetic fiber rope; therefore, it cannot be used in life safety situations.

_____

_____

8. ☐ T ☐ F   If during inspection of a braid-on-braid rope the sheath is found to be sliding on the core, the end of the rope should be cut, the excess sheath pulled off, and the end seared.

_____

_____

9. ☐ T ☐ F   Fifty percent of the kernmantle rope's strength lies within its core.

_____

_____

## MULTIPLE CHOICE

**Circle the letter before the most appropriate response.**

10. Firefighters A and B are discussing natural fiber rope. Firefighter A says that it is no longer acceptable to use natural fiber rope for utility purposes. Firefighter B says that natural fiber rope is no longer accepted for use in life safety applications. Who is correct?

    A. Firefighter A
    B. Firefighter B
    C. Both A and B
    D. Neither A nor B

11. Which of the following statements about manila rope is *not* correct?

    A. Manila rope should be soaked before it is used.
    B. Manila rope should be considered "used" if it is six months old and has not been employed.
    C. Manila rope is severely affected by chemicals and abrasion.
    D. Manila rope that is stored in a humid atmosphere loses one-half its strength in one year.

12. Which of the following statements about cotton rope is true?

    A. Its tensile strength is approximately equal to the tensile strength of sisal.
    B. Its tensile strength is considerably more than the tensile strength of manila.
    C. Its tensile strength is slightly more than the tensile strength of sisal.
    D. Its tensile strength is considerably less than the tensile strength of manila.

13. Firefighters A and B are discussing knots. Firefighter A says that a figure of eight on a bight can be tied either in the middle or at the end of a rope. Firefighter B says that this knot is preferred as the replacement for the bowline when using synthetic rope. Who is correct?

    A. Firefighter A
    B. Firefighter B
    C. Both A and B
    D. Neither A nor B

14. How should synthetic fiber rope be cleaned?

    A. With bleach in order to properly disinfect it
    B. According to the manufacturer's instructions
    C. By using a hoseline with a solid stream nozzle to thoroughly rinse away all debris
    D. All of the above

## IDENTIFICATION

**Identify the following items.**

15. NFPA 1983

    _____

    _____

16. Laid rope construction

    _____

    _____

17. Braided rope construction

    _____

    _____

18. Braid-on-braid rope construction

    _____

    _____

19. Kernmantle rope construction

    _____

    _____

**4**

**Identify the characteristics of the different types of synthetic rope. Write the correct letters in the blanks; some blanks may have more than one answer.**

20. _____ Has three to three and one-half times the tensile strength of manila rope

21. _____ Will float

22. _____ Not easily formed into knots and hitches

23. _____ Begins to lose strength at 300°F (149°C); begins to melt at temperatures from 450°F to 650°F (232°C to 343°C)

24. _____ Begins to lose strength at 200°F (93°C); begins to melt at temperatures from 285°F to 300°F (140°C to 149°C)

25. _____ Maintains approximately 75 to 80 percent of its strength when wet

26. _____ Has relatively high rate of deterioration from sunlight

27. _____ Resists abrasion

28. _____ Can be damaged by abrasion, so must be sheathed in another material to protect it from abrasion during its use

A. Nylon
B. Polypropylene
C. Polyethylene
D. Polyester
E. Kelvar® aramid fiber

**Identify the following rope formations by writing their correct names in the blanks.**

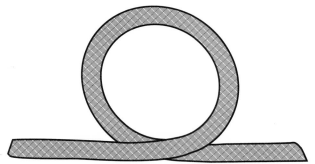

29. _____

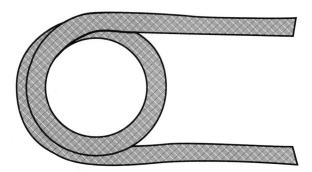

30. _____

**4**

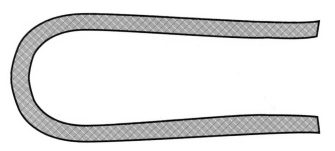

31. _____

**Identify each of the labeled parts as either the *running part*, the *working end*, or the *standing part*.**

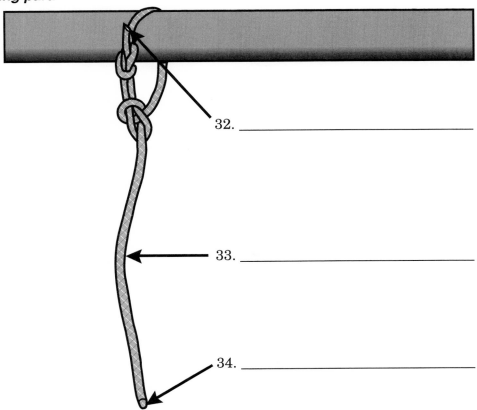

32. _____

33. _____

34. _____

**Identify the knots presented in the following illustrations, and give their common uses.**

35.  A.  Knot _____

     B.  Common use(s) _____

       _____

       _____

**4**

36. A. Knot _____

    B. Common use(s) _____

    _____

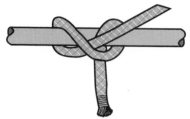

37. A. Knot _____

    B. Common use(s) _____

    _____

38. A. Knot _____

    B. Common use(s) _____

    _____

39. A. Knot _____

    B. Common use(s) _____

    _____

**4**

**Identify the proper knots and securing procedures for hoisting the following tools and equipment.**

40. Ladder _____

_____
_____
_____
_____
_____

41. Pike pole _____

_____
_____
_____
_____
_____

42. Axe _____

_____
_____
_____
_____
_____

43. Smoke ejector _____

_____
_____
_____
_____
_____

44. Dry hoseline _____

_____
_____
_____
_____
_____

**4**

45. Charged hoseline _____

_____

_____

_____

_____

_____

## LISTING

46. List the three principal ways of cleaning synthetic fiber rope.
    A. _____
    B. _____
    C. _____

47. List the two common methods of storing rope.
    A. _____
    B. _____

## SHORT ANSWER

**Briefly answer each question in your own words.**

48. What are the characteristics of synthetic fiber rope that make it desirable?

_____

_____

_____

49. How are static rope and dynamic rope different, and in what instances is each type used?

_____

_____

_____

_____

_____

_____

_____

_____

**4**

50. What is the purpose of using an overhand safety knot?

   _____

   _____

   _____

   _____

51. What is indicated by the presence of lumps in braid-on-braid rope?

   _____

52. Why is it important to carefully examine any type of damage or questionable wear to the sheath of a braid-on-braid rope?

   _____

   _____

53. What is the basic procedure for inspecting braided rope?

   _____

   _____

   _____

   _____

   _____

   _____

54. What is the basic inspection procedure for kernmantle rope?

   _____

   _____

   _____

   _____

   _____

   _____

55. Why can natural fiber rope not be cleaned effectively, and how is it cleaned?

   _____

   _____

   _____

   _____

   _____

**4**

56. In what ways can synthetic rope be dried?

_____

_____

_____

_____

_____

_____

57. What is the best method for storing kernmantle rope and other life safety rope? Why?

_____

_____

_____

_____

58. Each rope bag should have a label. What information should be included on this label?

_____

_____

59. What is the purpose of using a tag line?

_____

_____

**5**

# Rescue And Extrication

# Rescue And Extrication | 5

## DEFINITIONS OF KEY TERMS

**Define each of the following terms.**

1. Shoring

   _____

   _____

2. Cribbing

   _____

   _____

## TRUE/FALSE

**Mark each of the following statements true (T) or false (F). Correct each false statement.**

3. ☐ T ☐ F The combination spreader/shears powered hydraulic tool consists of two arms equipped with spreader tips that can be used for pushing only.

   _____

   _____

4. ☐ T ☐ F Manual hydraulic tools operate on the same principle as powered hydraulic tools.

   _____

   _____

5. ☐ T ☐ F There are two basic types of lifting bags: high pressure and low pressure.

   _____

   _____

6. ☐ T ☐ F The extremities carry is a four-person carry that is designed to be used for moving an unconscious victim.

   _____

   _____

**5**

7. ☐ T  ☐ F  The long backboard is one of the most common types of litters used by fire service personnel.

_____

_____

8. ☐ T  ☐ F  When moving a victim suspected of having a cervical spine injury, the rescuer supporting the victim's back should direct the other rescuers in their actions.

_____

_____

9. ☐ T  ☐ F  When preparing to lower a victim from a window, the ladder should be raised to a point just below the window where the rescue is to be made.

_____

_____

10. ☐ T  ☐ F  Except when the vehicle still has all of its wheels on the ground, stabilization of a vehicle involved in an accident is required to ensure maximum stability for extrication operations.

_____

_____

11. ☐ T  ☐ F  Different methods are used for freeing a victim trapped by the steering wheel on a rear-wheel drive car and for freeing a victim similarly trapped in a front-wheel drive car.

_____

_____

12. ☐ T  ☐ F  For cave rescues, it is recommended that rescuers carry two sources of light: a cap-mounted, battery-powered lantern and a waterproof flashlight.

_____

_____

13. ☐ T  ☐ F  Because the water temperature in all caves is usually around 55°F (13°C), hypothermia is a very possible danger to any victim or rescuer who may become immersed.

_____

_____

5

14. ☐ T ☐ F  The most successful approach to locating a victim within a cave is to first send small search teams on a rapid search through the most commonly traveled passages of the cave.

_____

_____

15. ☐ T ☐ F  All electrical wires or equipment and all victims in contact with such must be regarded as energized and dangerous.

_____

_____

16. ☐ T ☐ F  Making personal contact between the rescuer and victim in the water is a last-resort tactic.

_____

_____

17. ☐ T ☐ F  Hoistway doors of an elevator are equipped with a weight or spring system that applies a constant force against the door toward the open position.

_____

_____

18. ☐ T ☐ F  Only firefighters or trained rescuers should activate the elevator car's emergency stop switch.

_____

_____

## MULTIPLE CHOICE

**Circle the letter before the most appropriate response.**

19. Which of the following types of powered hydraulic tools is designed primarily for straight pushing operations but is also effective for pulling operations as well?
    A. Combination spreader/shears
    B. Spreader
    C. Shears
    D. Extension ram

20. Firefighters A and B are discussing powered hydraulic tools used for rescue. Firefighter A says that the cutting capability of the combination spreader/shears is greater than that of powered hydraulic shears. Firefighter B says that powered hydraulic shears can produce about 30,000 psi (206 850 kPa) of cutting force. Who is correct?
    A. Firefighter A
    B. Firefighter B
    C. Both A and B
    D. Neither A nor B

**5**

21. Which of the following statements about air lifting bags is *not* correct?
    A. Low-pressure bags have a greater lifting distance than do high-pressure bags.
    B. Low-pressure bags are considerably smaller than high-pressure bags.
    C. Low-pressure bags are most commonly used to lift or stabilize large vehicles or objects.
    D. Some high-pressure bags can lift up to 75 tons (68 040 kg).

22. In the presence of no life-threatening dangers, immobilizing a victim who is suspected of having a spinal injury on a long backboard requires _____ rescuer(s).
    A. One
    B. Two
    C. Three
    D. Four

23. Firefighters A and B are discussing procedures to follow when trapped in a burning building. Firefighter A says that a rescuer trapped in a burning building should try to retreat to the ground floor. Firefighter B says that if the rescuer is unable to go down and exit, the rescuer should go to a room that has a window opening to the street. Who is correct?
    A. Firefighter A
    B. Firefighter B
    C. Both A and B
    D. Neither A nor B

24. In many vehicle extrication instances, it is necessary for the rescuers to remove the vehicle's roof. How is this accomplished?
    A. By cutting all the roof posts and removing the roof entirely
    B. By cutting only the front posts and folding the roof back over the trunk
    C. Neither A nor B
    D. Either A or B

25. When a victim is trapped by the steering wheel or the dashboard in a rear-wheel drive vehicle, how should the rescuers free the victim?
    A. By pulling the steering column
    B. By displacing the dashboard
    C. By cutting away the floorboard
    D. None of the above

26. Which of the following statements about trench cave-in rescue is *not* correct?
    A. Do not use heavy equipment until the exact location and number of victims are known.
    B. Use picks to quickly uncover and locate the victim.
    C. Use shoring or cribbing to hold back weakened earth formations.
    D. Remove debris to a clear area.

27. Firefighters A and B are discussing electrical emergencies. Firefighter A says that appropriately trained fire department personnel, if using standard bolt cutters, may cut or move energized wires. Firefighter B says that energized electrical wires should never be cut because the wires may recoil or "jump" after being cut. Who is correct?
   A. Firefighter A
   B. Firefighter B
   C. Both A and B
   D. Neither A nor B

28. Firefighters A and B are discussing elevator rescues. Firefighter A says that exits are usually located on the top and at the sides of an elevator car. Firefighter B says that the side exit panels are normally locked from outside the car and that the top exit panel may be opened from the inside with a special key. Who is correct?
   A. Firefighter A
   B. Firefighter B
   C. Both A and B
   D. Neither A nor B

29. Which of the following statements about controlling an elevator is *not* correct?
   A. Independent service requires a key to unlock the panel that protects the toggle switch.
   B. Emergency service requires a key to activate the system.
   C. Movement and direction of travel of a car operating in the independent service mode are controlled by a keyed switch.
   D. A car in the emergency service mode of operation will respond only to signals initiated from within the car.

**MATCHING**

**Match the following harnesses with their correct descriptions. Write the correct letters in the blanks.**

30. _____ Sit-type harness that is designed to support two people and that has additional support over the shoulders to prevent the wearer from becoming inverted on the rope

31. _____ Sit-type harness designed to support the weight of two people (victim and rescuer)

32. _____ Harness that goes around the wearer's waist and that is used only to secure the firefighter to a ladder or other objects

A. Class I harness
B. Class II harness
C. Class III harness

**5**

**Identify the following items.**

33. Extrication incidents

_____

_____

34. Rescue incidents

_____

_____

35. Block

_____

_____

36. Tackle

_____

_____

37. Simple tackle

_____

_____

38. Compound tackle

_____

_____

39. NFPA 1983

_____

_____

40. Primary search

_____

_____

41. Secondary search

_____

_____

**Identify the parts of the block and tackle illustrated. Write the correct names in the blanks**.

Fall line           Standing block
Running block     Leading block

43. _____

44. _____

42. _____

45. _____

**Identify the following types of structural collapse by giving the description of each type.**

46. Lean-to collapse _____

_____

_____

_____

_____

_____

47. Pancake collapse _____

_____

_____

_____

_____

_____

**5**

48. V-type collapse _____

_____

_____

_____

_____

_____

**Provide the requested information.**

49. Identify the three types of elevator keys, and explain how each is used.

    A. _____

       _____

    B. _____

       _____

    C. _____

       _____

## LISTING

50. List the four basic types of powered hydraulic tools used in the rescue service.

    A. _____

    B. _____

    C. _____

    D. _____

51. List four sources of compressed air that can be used to power pneumatic tools.

    A. _____

    B. _____

    C. _____

    D. _____

52. List at least eight safety rules to follow when using air bags.

_____

_____

_____

_____

_____

_____

_____

_____

_____

_____

_____

**5**

53. List at least five safety rules to follow when using block and tackle.

_____

_____

_____

_____

_____

_____

_____

_____

_____

54. List at least eight guidelines that the firefighter should follow to help ensure a swift and safe rescue operation once the firefighter enters the involved structure.

_____

_____

_____

_____

_____

_____

_____

_____

_____

_____

_____

_____

_____

_____

_____

_____

55. List the three methods in general of gaining access to victims trapped within a vehicle.

A. _____

B. _____

C. _____

**5**

56. At an extrication incident, assessment should begin as soon as the first emergency vehicle *approaches* the accident scene (the initial size-up). List the factors rescue personnel should try to determine as they approach.

_____

_____

_____

_____

_____

_____

_____

57. Once at the scene of an extrication incident, rescue personnel should assess the immediate area around each vehicle and assess the entire scene in more detail. List the factors rescue personnel should consider during this assessment.

_____

_____

_____

_____

_____

_____

_____

_____

_____

_____

_____

_____

58. List the three items used most frequently to prevent an automobile from moving vertically.

A. _____

B. _____

C. _____

59. List five safety precautions that firefighters and officers must remember when they are involved in cave-ins and excavation rescues.

A. _____

_____

_____

_____

_____

**5**

B. _____
_____
_____
_____
_____
_____
_____

C. _____
_____
_____
_____
_____
_____
_____
_____

D. _____
_____
_____
_____
_____
_____
_____
_____

E. _____
_____
_____
_____
_____
_____
_____

60. List five methods that can be used to rescue a victim during a water emergency.

A. _____

B. _____

C. _____

D. _____

E. _____

**5**

61. List in order the steps for handling an industrial extrication.

Step 1 _____

Step 2 _____

Step 3 _____

Step 4 _____

Step 5 _____

Step 6 _____

Step 7 _____

Step 8 _____

62. List factors rescue personnel should consider when surveying the situation during an industrial extrication.

_____

_____

_____

_____

_____

_____

_____

## SHORT ANSWER

**Briefly answer each question in your own words.**

63. What are the primary advantage and disadvantage of the porta-power compared to the hydraulic jack?

_____

_____

_____

_____

_____

_____

_____

_____

_____

64. In what type of situations is the hydraulic jack commonly used?

_____

_____

_____

**5**

65. The pneumatic chisel can be especially effective for auto extrication. How is it used in such situations?

_____

_____

_____

66. The firefighter should exercise caution when operating a pneumatic chisel in an area containing hazardous atmospheres. Why?

_____

_____

_____

67. For what purposes are leak-sealing air bags used?

_____

_____

_____

68. When should life safety harnesses be inspected?

_____

_____

_____

69. What is the purpose of using a guideline when lowering a victim?

_____

_____

_____

70. If a firefighter is rescuing a downed firefighter whose air supply is deleted, should the firefighter with air remaining share her or his air supply? If so, how?

_____

_____

_____

_____

_____

_____

_____

_____

_____

**5**

71. Following the assessment at the scene of a vehicle accident, the rescuers must stabilize the vehicle(s). What does this mean?

_____

_____

_____

_____

_____

_____

72. When tunneling is used to reach a void under a floor, the tunnel should be driven along a wall whenever possible. Why?

_____

_____

73. At a trench cave-in, the most common problem that faces the victims is the reduced ability to breathe. What are some of the methods of assisting the victim's breathing?

_____

_____

_____

_____

_____

_____

_____

_____

74. In what ways can rescuers most safely traverse unstable ice in order to effect a rescue?

_____

_____

_____

_____

_____

_____

# Forcible
# Entry

# Forcible Entry | 6

## DEFINITIONS OF KEY TERMS

**Define each of the following terms.**

1. Doorjamb

   _____

   _____

2. Rabbeted jamb

   _____

   _____

3. Stopped jamb

   _____

   _____

4. Tempered plate glass

   _____

   _____

5. Lexan® plastic

   _____

   _____

## TRUE/FALSE

**Mark each of the following statements true (T) or false (F). Correct each false statement.**

6. ☐ T ☐ F  Generally, handsaws are used for forcible entry, and power saws are used for rescue.

   _____

   _____

7. ☐ T ☐ F  The reciprocating saw is a highly controllable saw that is well suited for cutting wood or metal.

   _____

   _____

**6**

8. ☐ T ☐ F    It is not acceptable to use a cheater bar with a prying tool in order to provide additional leverage.

_____

_____

9. ☐ T ☐ F    Striking tools are the most common and basic hand tools.

_____

_____

10. ☐ T ☐ F    Residential swinging doors generally open outward, and those in public buildings should open inward.

_____

_____

11. ☐ T ☐ F    It is generally considered impractical to force swinging metal doors.

_____

_____

12. ☐ T ☐ F    Fire doors are used primarily to protect openings in division walls and walls of vertical shafts.

_____

_____

13. ☐ T ☐ F    Self-closing fire doors normally remain open but close when heat actuates the closing device.

_____

_____

14. ☐ T ☐ F    The most effective entry through Lexan is by striking it with a pick-head axe.

_____

_____

15. ☐ T ☐ F    Generally, the floors of upper stories of family dwellings are concrete slab.

_____

_____

16. ☐ T ☐ F    When cutting open a metal wall, the firefighter should cut in an area midway between the studs or supports to avoid any unnecessary structural damage.

_____

_____

## MULTIPLE CHOICE

**Circle the letter before the most appropriate response.**

17. Which of the following statements about cutting and prying tools is *not* correct?
    A. Most cutting tools are designed to cut a variety of materials.
    B. Cutting tools may be either manual or power tools.
    C. Most cutting tools are designed to cut only specific types of materials.
    D. Prying tools may be either manual or power tools.

18. Firefighters A and B are discussing axes. Firefighter A says that the flat-head axe is designed primarily for cutting but that it is also often used as a striking tool. Firefighter B says that the blades of pick-head axes make them adaptable for prying operations. Who is correct?
    A. Firefighter A
    B. Firefighter B
    C. Both A and B
    D. Neither A nor B

19. Which of the following tools would be the most suitable for cutting a heavy metal enclosure within a nonflammable atmosphere?
    A. Flat-head axe
    B. Rotary rescue saw
    C. Powered hydraulic shears
    D. Oxyacetylene cutting torch

20. Which of the following tools is specifically designed for prying inward-swinging doors mounted in metal frames?
    A. Rabbit tool
    B. Halligan tool
    C. Kelly tool
    D. Crowbar

21. The three general types of wooden swinging doors are _____.
    A. Ledge, tubular, and slab
    B. Ledge, hollow core, and solid core
    C. Ledge, tubular, and panel
    D. Ledge, panel, and slab

22. Metal swinging doors may be classified as _____.
    A. Hollow metal, hollow core, and solid metal
    B. Hollow metal, metal covered, and tubular
    C. Hollow metal, tubular, and solid metal covered
    D. Tubular, ledge, and metal covered

**6**

23. Firefighter A says that before attempting to force any door, the firefighter should check to see whether the door is locked. Firefighter B agrees and says that if the door is locked, the firefighter should check to see whether or not the hinge pins can be removed. Who is correct?
    A. Firefighter A
    B. Firefighter B
    C. Both A and B
    D. Neither A nor B

24. Which of the following may be used to pull or force a cylinder lock?
    A. A-tool
    B. Key tool
    C. K-tool
    D. Both A and C

25. Which of the following statements about tempered plate glass is *not* correct?
    A. Tempered plate glass door panels are considerably less expensive than other glass-paneled doors of similar size.
    B. A tempered plate glass door that is locked is almost impossible to spring with forcible entry tools.
    C. Tempered plate glass will withstand, without breaking, a temperature of 650°F (343°C) on one side while the other side is exposed to ordinary atmospheric temperature.
    D. Forcing entry through a tempered plate glass door should be the last available means to gain entrance.

26. _____ overhead doors are best accessed by cutting a triangle-shaped opening large enough through which firefighters can crawl.
    A. Remote-controlled sectional
    B. Folding
    C. Rolling steel
    D. Slab

27. Firefighters A and B are discussing forcing overhead doors. Firefighter A says that such doors cannot be forced and that a panel must be knocked out so that the door latch can be turned from the inside. Firefighter B says that with overhead slab doors, it sometimes is possible to pry outward with a bar at each side near the bottom so that the lock bar will bend enough to pass the keeper. Who is correct?
    A. Firefighter A
    B. Firefighter B
    C. Both A and B
    D. Neither A nor B

28. Which of the following types of windows is the most difficult to force?
    A. Casement
    B. Jalousie
    C. Checkrail
    D. Projected

6

29. Wooden joists of wood floor construction are usually spaced a maximum of
    _____ apart.
    A.  12 inches (300 mm)
    B.  16 inches (400 mm)
    C.  18 inches (450 mm)
    D.  24 inches (600 mm)

30. Firefighters A and B are discussing cutting through wood floors. Firefighter A says
    that carpeting should be removed before a floor is cut. Firefighter B agrees but says
    that other covering such as tile or linoleum does not have to be removed before
    cutting through the floor. Who is correct?
    A.  Firefighter A
    B.  Firefighter B
    C.  Both A and B
    D.  Neither A nor B

31. Which of the following statements about breaching masonry and veneered walls is
    *not* correct?
    A.  Charged hoselines should be in position before breaching a wall at a fire.
    B.  Masonry walls can be the toughest type to breach.
    C.  A battering ram may be used when breaching masonry walls.
    D.  Air chisels or hydraulic spreaders are better suited for breaching masonry walls
        than are rotary rescue saws.

## IDENTIFICATION

**Identify the following items.**

32. Hollow core door

    _____

    _____

33. Solid core door

    _____

    _____

34. Rapid entry key box

    _____

    _____

35. Class A openings

    _____

    _____

**6**

36. Class B openings

_____

_____

37. Penetrating nozzle

_____

_____

**Identify the following types of tools by writing the correct letters in the blanks. Some blanks may have more than one correct answer.**

CU — Cutting tool
ST — Striking tool
PR — Prying tool
PU — Pushing/pulling tool

38. _____ Rotary rescue saw

39. _____ Rabbit tool

40. _____ Powered hydraulic extension ram

41. _____ Sledgehammer

42. _____ Axe

43. _____ Pike pole

44. _____ Picks

45. _____ Spanner wrench

46. _____ Halligan tool

47. _____ Handsaw

48. _____ Powered hydraulic spreader

49. _____ Bolt cutters

50. _____ Punches

51. _____ Oxyacetylene cutting torch

52. _____ Plaster hook

53. _____ Hammer

54. _____ Kelly tool

**Identify the procedures for forcible entry tool care for the following tools and tool parts.**

55. Wooden handles _____

_____

_____

_____

_____

_____

**6**

56. Fiberglass handles _____

_____

_____

_____

_____

57. Cutting edges _____

_____

_____

_____

_____

_____

58. Plated surfaces _____

_____

_____

_____

_____

_____

59. Unprotected metal surfaces _____

_____

_____

_____

_____

_____

60. Power equipment _____

_____

_____

_____

_____

_____

**6**

**Identify the procedures used to collapse the mechanisms which hold the following types of revolving doors in place.**

61. Panic-proof type _____
_____
_____
_____
_____

62. Drop-arm type _____
_____
_____
_____
_____

63. Metal-braced type _____
_____
_____
_____
_____

**Identify the methods for forcing entry through the following types of windows.**

64. Wooden checkrail _____
_____
_____
_____
_____

65. Metal checkrail _____
_____
_____
_____
_____

66. Casement _____
_____
_____
_____
_____

**6**

67. Projected _____

_____

_____

_____

68. Awning or jalousie _____

_____

_____

_____

_____

## LISTING

69. List the six safety rules to follow when using power saws.

A. _____

_____

B. _____

_____

C. _____

_____

D. _____

_____

E. _____

_____

F. _____

70. List safety rules to observe when using oxyacetylene cutting equipment.

_____

_____

_____

_____

_____

_____

_____

_____

**6**

71. List the three factors to consider when determining which method should be used to force a swinging door.

   A. _____

   B. _____

   C. _____

72. List the four ways to determine whether partitions contain fire.

   A. _____

   B. _____

   C. _____

   D. _____

## SHORT ANSWER

**Briefly answer each question in your own words.**

73. What two things may occur when a tool is pushed beyond the limits of its design and purpose?

   _____

   _____

74. Why must the use of power saws in flammable atmospheres be prohibited?

   _____

   _____

75. Hand prying tools use leverage to provide a mechanical advantage. What does this mean?

   _____

   _____

76. Why is it important to maintain the blade of an axe at the proper thickness?

   _____

   _____

   _____

   _____

   _____

**6**

77. What is the basic procedure for breaking glass?

_____

_____

_____

_____

_____

_____

_____

_____

78. What is the basic procedure for forcing a swinging door that opens toward the firefighter?

_____

_____

_____

_____

_____

_____

_____

_____

79. How is the procedure for forcing a swinging door with a stopped jamb that opens away from the firefighter different from the procedure for forcing one with a rabbeted jamb?

_____

_____

_____

_____

_____

80. How can sliding patio doors be forced, and what are the devices sometimes in place that make forcing difficult?

_____

_____

_____

_____

_____

_____

**6**

81. What precautionary measure should firefighters take when passing through an opening protected by a fire door? Why?

_____

_____

_____

_____

82. What is the most feasible way of opening a concrete floor?

_____

_____

83. What is the basic procedure for opening wood frame walls?

_____

_____

_____

_____

**7**

# Ventilation

# Ventilation | 7

## DEFINITIONS OF KEY TERMS

**Define each of the following terms**.

1. Ventilation

   _____

   _____

2. Mushrooming

   _____

   _____

3. Flashover

   _____

   _____

4. Backdraft

   _____

   _____

5. Vertical ventilation

   _____

   _____

6. Stack effect

   _____

   _____

7. Horizontal ventilation

   _____

   _____

**7**

**Mark each of the following statements true (T) or false (F). Correct each false statement.**

8. ☐ T ☐ F  Building type and design are the initial factors to consider in determining whether to use horizontal or vertical ventilation.

_____

_____

9. ☐ T ☐ F  Slate and tile roofs must be opened by using a rescue saw with a masonry-cutting blade.

_____

_____

10. ☐ T ☐ F  Tin roofs can be sliced open and peeled back with tin snips.

_____

_____

11. ☐ T ☐ F  With an arched roof, a hole of considerable size may be cut or burned through the network sheathing and roofing any place without causing collapse of the roof structure.

_____

_____

12. ☐ T ☐ F  Especially in buildings of balloon-frame construction, the first extension from a basement fire will commonly be into the attic.

_____

_____

13. ☐ T ☐ F  For hydraulic ventilation, the fog stream should be set on a wide fog pattern that will cover 95 to 100 percent of the window or door opening from which the smoke will be pushed.

_____

_____

14. ☐ T ☐ F  Because an HVAC system may draw heat and smoke into the duct before it is shut down, firefighters should always check around the ductwork for fire extension during overhaul.

_____

_____

**MULTIPLE CHOICE**

**Circle the letter before the most appropriate response.**

15. Each of the following types of existing roof openings generally provides an adequate opening for ventilation purposes *except* _____.
    A. Monitors
    B. Skylights
    C. Scuttle hatches
    D. Stairway doors

16. When opening a flat wood joist roof, what is the next step once the location of roof supports have been marked?
    A. Cut the wood decking diagonally alongside the joist toward the hole.
    B. Cut through the two joists adjacent to the opening location.
    C. Remove the built-up roof material.
    D. Pry up the roof boards.

17. Firefighters A and B are discussing opening arched roofs. Firefighter A says that the procedure for opening an arched roof is basically the same as for flat or pitched roofs except that a roof ladder cannot always be used on an arched roof. Firefighter B says that firefighters should work from an aerial ladder or platform extended to the roof. Who is correct?
    A. Firefighter A
    B. Firefighter B
    C. Both A and B
    D. Neither A nor B

18. Firefighters A and B are discussing concrete roofs. Firefighter A says that all concrete roofs are extremely difficult to break through and that opening them should be avoided whenever possible. Firefighter B says that lightweight concrete slabs are relatively easy to penetrate and that most of them can be penetrated with a hammer-head pick, power saw with concrete blade, or other penetrating tool. Who is correct?
    A. Firefighter A
    B. Firefighter B
    C. Both A and B
    D. Neither A nor B

19. Which of the following types of structures lends itself to the application of horizontal ventilation?
    A. Residential-type buildings in which the fire has not involved the attic area
    B. Buildings with large, unsupported open spaces under the roof
    C. Multistoried structures in which the floors involved in fire are below the attic or below the top floor if there is no attic
    D. All of the above

**7**

**Identify the following items.**

20. Trench ventilation

_____

_____

21. Strip ventilation

_____

_____

22. Mechanical ventilation

_____

_____

23. Hydraulic ventilation

_____

_____

**Identify the following common roof styles.**

24. _____

25. _____

26. _____

7

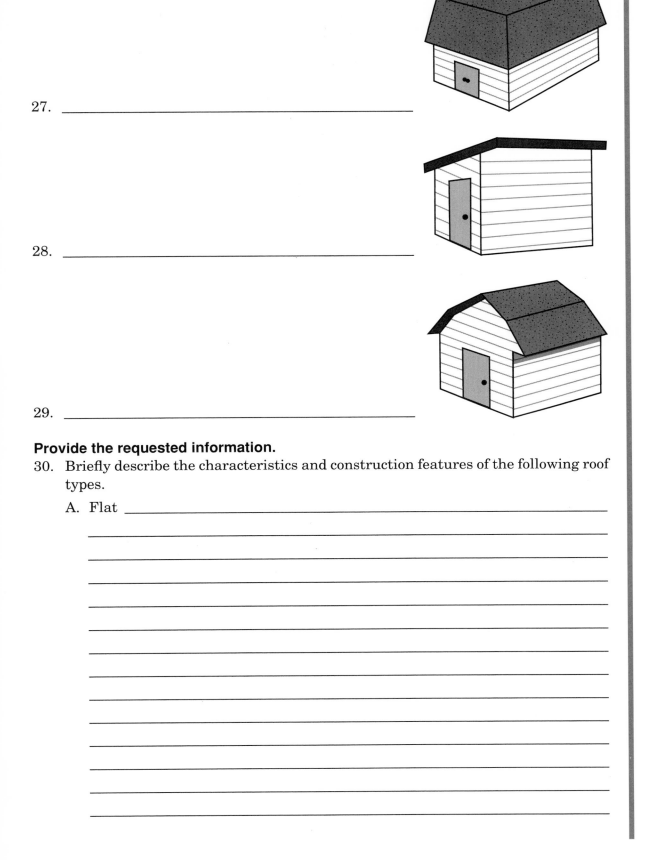

27. _____

28. _____

29. _____

**Provide the requested information.**

30. Briefly describe the characteristics and construction features of the following roof types.

    A. Flat _____

    _____

    _____

    _____

    _____

    _____

    _____

    _____

    _____

    _____

    _____

    _____

    _____

**7**

B. Pitched _____
_____
_____
_____
_____
_____
_____
_____
_____
_____
_____

C. Arched _____
_____
_____
_____
_____
_____
_____
_____
_____
_____
_____

31. Given the wind direction as shown, identify the *leeward* side and *windward* side of the structure in the following illustration.

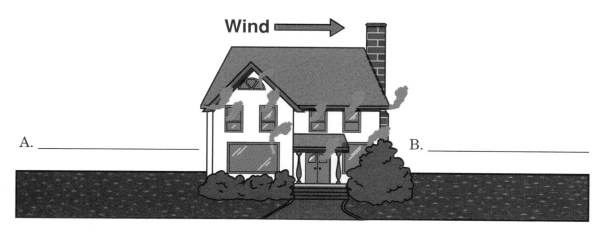

A. _____     B. _____

**LISTING**

32. List the advantages of ventilation.

_____

_____

_____

_____

_____

_____

_____

_____

_____

_____

_____

_____

_____

_____

_____

_____

_____

_____

_____

33. What three questions should be considered when making ventilation decisions?

A. _____

B. _____

C. _____

34. List the factors concerning the building involved that must be considered when determining whether to use horizontal or vertical ventilation.

_____

_____

_____

_____

_____

_____

_____

**7**

35. List at least eight safety precautions that should be practiced when performing vertical ventilation.

_____

_____

_____

_____

_____

_____

_____

_____

_____

_____

_____

_____

_____

_____

36. List the warning signs of an unsafe roof condition.

_____

_____

_____

_____

_____

_____

37. List common factors that can destroy the effectiveness of vertical ventilation.

_____

_____

_____

_____

_____

38. List ways in which horizontal ventilation may be upset.

_____

_____

_____

_____

_____

39. List advantages and disadvantages of forced ventilation.

_____

_____

_____

_____

_____

_____

40. List at least four advantages of positive-pressure ventilation.

_____

_____

_____

_____

_____

_____

_____

_____

## SHORT ANSWER

**Briefly answer each question in your own words.**

41. When performing top ventilation in a high-rise building, why is it important that the door at the top of the stair shaft be removed?

_____

_____

_____

_____

42. What may be the result of making a ventilation opening before the fire is located?

_____

_____

_____

_____

43. Why is using existing openings, if possible, for vertical ventilation purposes beneficial, and what must firefighters realize about the size and location of such openings?

_____

_____

_____

_____

**7**

44. What is the basic procedure for opening a flat or pitched roof constructed with wood joists or rafters?

_____

_____

_____

_____

_____

_____

_____

_____

45. What may be the result if fire streams are projected downward through a ventilation opening?

_____

_____

_____

_____

_____

46. What is negative-pressure ventilation, and how is it accomplished?

_____

_____

_____

_____

_____

_____

_____

47. What is positive-pressure ventilation, and how is it accomplished?

_____

_____

_____

_____

_____

_____

_____

8

**Ladders**

# Ladders | **8**

## DEFINITIONS OF KEY TERMS

**Define each of the following terms.**

1. Base section

   _____
   _____

2. Beam

   _____
   _____

3. Beam bolts

   _____
   _____

4. Butt

   _____
   _____

5. Butt spurs

   _____
   _____

6. Fly

   _____
   _____

7. Guides

   _____
   _____

8. Halyard

   _____
   _____

**8**

9. Heat sensor label

_____

_____

10. Hooks

_____

_____

11. Pawls

_____

_____

12. Protection plates

_____

_____

13. Pulley

_____

_____

14. Rails

_____

_____

15. Rungs

_____

_____

16. Safety shoes

_____

_____

17. Spurs

_____

_____

18. Staypoles

_____

_____

8

19. Stops

_____

_____

20. Tie rods

_____

_____

21. Toggle

_____

_____

22. Top or tip

_____

_____

23. Truss block

_____

_____

## TRUE/FALSE

**Mark each of the following statements true (T) or false (F). Correct each false statement.**

24. ☐ T ☐ F  The main uses of aerial ladders are rescue, ventilation, elevated master stream application, and gaining access to upper levels.

_____

_____

25. ☐ T ☐ F  One difference in telescoping aerial platform apparatus and aerial ladder platform apparatus is that a telescoping aerial platform apparatus is designed with a large ladder that allows firefighters to routinely climb back and forth from the platform.

_____

_____

26. ☐ T ☐ F  Telescoping aerial platforms are equipped with built-in piping and nozzles for providing elevated streams.

_____

_____

**8**

27. ☐ T ☐ F   Clear, straight-grained, coast Douglas fir is a favored wood for ladder beams because it is relatively free of knots and pitch pockets.

_____

_____

28. ☐ T ☐ F   Firefighters should remember to use leg muscles — not back muscles — when lifting ladders.

_____

_____

29. ☐ T ☐ F   A residential story will average 10 to 12 feet (3 m to 4 m) from floor to floor, with a 3-foot (1 m) distance from the floor to the windowsill; stories of commercial buildings will average 14 feet (4.3 m) from floor to floor, with a 4-foot (1.3 m) distance from the floor to the windowsill.

_____

_____

30. ☐ T ☐ F   The distance of the butt of the ladder from the building establishes the angle formed by the ladder and the ground; an angle of 75½ degrees is the desired angle.

_____

_____

31. ☐ T ☐ F   Whenever possible, a ladder should be extended before it is pivoted.

_____

_____

32. ☐ T ☐ F   Typically, ladders of 35 feet (11 m) or larger should be raised by at least four firefighters.

_____

_____

## MULTIPLE CHOICE

**Circle the letter before the most appropriate response.**

33. Which of the following statements about ladder construction is correct?
    A. With wooden ladders, truss construction will be lighter than solid beams.
    B. For ladders with lengths over 24 feet (8 m), truss construction provides the lightest weight ladder.
    C. Ladder rungs must not be less that 1 inch (25 mm) in diameter and spaced on 12-inch (300 mm) centers.
    D. Fire service ladders may be constructed only of either metal or wood.

34. Firefighters A and B are discussing ladder service testing. Firefighter A says that strength testing should be performed on all single, roof, folding, and extension ladders. Firefighter B says that the procedures for strength testing extension and single ladders differs from the procedures for strength testing folding ladders. Who is correct?
    A. Firefighter A
    B. Firefighter B
    C. Both A and B
    D. Neither A nor B

35. When three or more firefighters are carrying an extension ladder, which firefighter should give the appropriate commands?
    A. The firefighter at the tip
    B. The firefighter at the butt
    C. A firefighter near the center of the ladder
    D. The firefighter at either end

36. When preparing for a three-firefighter flat-shoulder carry from the ground, how should the firefighters be positioned around the ladder?
    A. All on the same side of the ladder, facing the butt
    B. All on the same side of the ladder, facing the tip
    C. Two on the same side of the ladder but on either end and one on the opposite side near the midpoint of the ladder, facing the butt
    D. Two on the same side of the ladder but on either end and one on the opposite side near the midpoint of the ladder, facing the tip

37. When preparing for a four-firefighter flat arm's length carry, how should the firefighters be positioned around the ladder?
    A. Two on the same side of the ladder but on either end and two on the opposite side equally spaced from the ends, facing the butt
    B. Two on the same side of the ladder but on either end and two, one on either side of the ladder, near the midpoint of the ladder, facing the tip
    C. All equally spaced along the same side of the ladder, facing the butt
    D. Two on either side at both ends of the ladder, facing the butt

38. Firefighters A and B are discussing the raising and placement of ladders. Firefighter A says that the presence of energized wires must be considered when placing aluminum ladders. Firefighter B says that the same consideration should be made when placing wooden or fiberglass ladders. Who is correct?
    A. Firefighter A
    B. Firefighter B
    C. Both A and B
    D. Neither A nor B

**8**

39. Currently all North American manufacturers of metal and fiberglass ground ladders require that their ladders be in the _____ position; however, most wooden ground ladders should be positioned with the _____.
   A. Fly in; fly out
   B. Fly in; fly in
   C. Fly out; fly in
   D. Fly out; fly out

40. Which of the following statements about raising ladders with three firefighters is correct?
   A. One firefighter heels the ladder as the other two firefighters at the ladder tip advance.
   B. Two firefighters heel the ladder as the other firefighter at the ladder tip advances.
   C. One firefighter heels the ladder as one firefighter steadies the ladder at midpoint and as the other firefighter at the tip advances.
   D. None of the above are correct.

## IDENTIFICATION

**Identify the following items.**
41. NFPA 1931

_____

_____

42. NFPA 1932

_____

_____

43. NFPA 1901

_____

_____

44. NFPA 1904

_____

_____

45. Aerial apparatus

_____

_____

46. Ground ladder

_____

_____

**8**

47. Single ladder

    _____

    _____

48. Roof ladder

    _____

    _____

49. Folding ladder

    _____

    _____

50. Extension ladder

    _____

    _____

51. Pole ladder

    _____

    _____

52. Combination ladder

    _____

    _____

53. Pompier ladder

    _____

    _____

**Identify the types of ladders illustrated, and identify the total weight load specified by NFPA for which each must be rated and the use of each.**

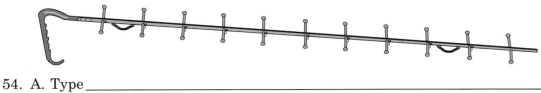

54. A. Type _____

    B. Weight load _____

    C. Use _____

    _____

    _____

    _____

**8**

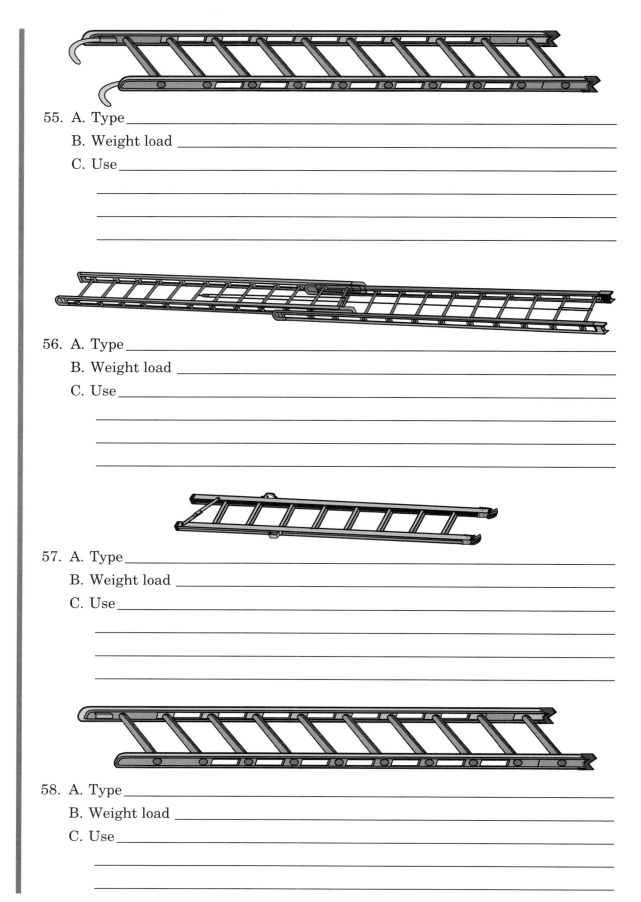

55. A. Type_____

    B. Weight load _____

    C. Use_____

    _____

    _____

    _____

56. A. Type_____

    B. Weight load _____

    C. Use_____

    _____

    _____

    _____

57. A. Type_____

    B. Weight load _____

    C. Use_____

    _____

    _____

    _____

58. A. Type_____

    B. Weight load _____

    C. Use_____

    _____

**8**

**Provide the requested information.**

59. Identify the inspection and maintenance procedures for wooden, metal, and fiberglass ladders.

_____
_____
_____
_____
_____
_____
_____
_____
_____
_____
_____
_____

60. Briefly describe the procedures for performing the following service tests for ladders.

A. Horizontal bending test _____

_____
_____
_____
_____
_____
_____
_____
_____
_____
_____

**8**

B.  Roof ladder hook test _____

_____

_____

_____

_____

_____

_____

_____

_____

61.  Identify the proper ladder placement for the following ladder uses.

A.  To effect ventilation from a window _____

_____

_____

_____

B.  To reach the roof _____

_____

_____

_____

C.  To use as a vantage point from which to direct a hose stream into a window opening when no entry is to be made _____

_____

_____

_____

D.  To use for rescue from a window _____

_____

_____

_____

_____

**8**

**LISTING**

62. List advantages and disadvantages of metal fire service ladders.

_____

_____

_____

_____

_____

_____

_____

_____

63. List advantages and disadvantages of fiberglass fire service ladders.

_____

_____

_____

_____

_____

_____

64. List the instances when fire service ladders should be service tested as required by NFPA 1932.

_____

_____

_____

_____

_____

_____

_____

65. List all the ladders that, according to NFPA 1904, aerial apparatus are required to carry.

_____

_____

_____

_____

_____

**8**

66. List at least eight ladder safety rules.

_____
_____
_____
_____
_____
_____
_____
_____
_____
_____
_____
_____
_____

## SHORT ANSWER

**Briefly answer each question in your own words.**

67. How can firefighters tell that a new ladder meets NFPA 1931?

_____
_____
_____
_____

68. When and how should ladders be cleaned?

_____
_____
_____
_____
_____

69. When should fire department ladders be inspected and tested?

_____
_____
_____
_____

**8**

70. What should be done with a ladder that has failed service testing?

_____

_____

71. NFPA 1901 requires fire department pumpers to be equipped with a certain complement of ladders. What are these ladders, and how are they usually mounted?

_____

_____

_____

_____

72. When removing a ladder from an apparatus mount for a one-person low-shoulder carry, toward which end of the ladder should the firefighter face?

_____

_____

73. Which end of the ladder should firefighters face when preparing for a two-firefighter low-shoulder carry from the ground?

_____

_____

74. With a one-firefighter single ladder raise, where should the butt of the ladder be placed?

_____

_____

_____

75. What are the two basic ways for two firefighters to raise a ladder, and who determines which one to use?

_____

_____

_____

76. How are firefighters positioned for a four-firefighter ladder raise?

_____

_____

77. Once a ladder has been raised, it should be secured. How is this accomplished?

_____

_____

**8**

78. How should a firefighter carry a tool, which can be carried in one hand, up and down a ladder?

_____

_____

_____

_____

79. How can the firefighter be secured to a ground ladder so that both hands of the firefighter are free to use on other tasks?

_____

_____

80. How does a firefighter assist a conscious victim down a ladder?

_____

_____

_____

_____

_____

_____

81. In what ways can a firefighter bring an unconscious victim down a ladder without entangling the victim's limbs in the rungs?

_____

_____

_____

_____

_____

_____

*Courtesy of Tom McCarthy, Chicago, IL.*

# Water Supply

# Water Supply | 9

## DEFINITIONS OF KEY TERMS

**Define each of the following terms.**

1. Circulating feed

_____

_____

2. Pressure

_____

_____

## TRUE/FALSE

**Mark each of the following statements true (T) or false (F). Correct each false statement.**

3. ☐ T ☐ F  Water distribution system valves should be operated every two years to keep them in good condition.

_____

_____

4. ☐ T ☐ F  One of the common types of indicator valves is the post indicator valve (PIV).

_____

_____

5. ☐ T ☐ F  Wet barrel hydrants may only be used in areas that do not have freezing weather; dry barrel hydrants are used in climates where freezing is expected.

_____

_____

**9**

6. ☐ T ☐ F   Although the installation of fire hydrants is usually the responsibility of the fire chief or fire marshal, the location, spacing, and distribution of fire hydrants should be the responsibility of water department personnel.

_____

_____

7. ☐ T ☐ F   During a relay pumping operation, a water supply officer must be appointed to determine the distance between pumpers and to coordinate water supplies.

_____

_____

## MULTIPLE CHOICE

**Circle the letter before the most appropriate response.**

8. Firefighters A and B are discussing water distribution system valves. Firefighter A says that the OS&Y (outside screw and yoke) valve is one type of nonindicating valve that is used. Firefighter B says that nonindicating valves are normally buried or installed in manholes. Who is correct?
   A. Firefighter A
   B. Firefighter B
   C. Both A and B
   D. Neither A nor B

9. Firefighters A and B are discussing water shuttle operations. Firefighter A says that apparatus attacking the fire may draft directly from the portable tanks. Firefighter B says that other apparatus may draft water from the tanks and then supply the attack apparatus. Who is correct?
   A. Firefighter A
   B. Firefighter B
   C. Both A and B
   D. Neither A nor B

## IDENTIFICATION

**Identify the following items.**

10. Direct pumping system

_____

_____

11. Gravity system

_____

_____

12. Combination system

_____

_____

13. Dead-end hydrant

_____

_____

14. Indicating valve

_____

_____

15. Flow pressure

_____

_____

16. Static pressure

_____

_____

17. Normal operating pressure

_____

_____

18. Residual pressure

_____

_____

19. Pitot tube

_____

_____

20. NFPA 1903

_____

_____

**9**

Identify the water lines of the following water distribution system as either *primary feeders, secondary feeders, or distributors.* Write the correct names in the blanks.

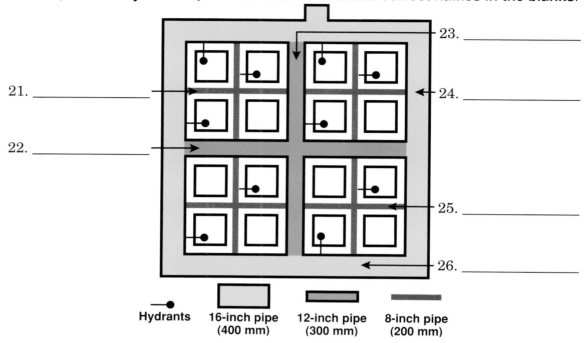

21. _____

22. _____

23. _____

24. _____

25. _____

26. _____

Hydrants    16-inch pipe (400 mm)    12-inch pipe (300 mm)    8-inch pipe (200 mm)

**Provide the requested information.**

27. Identify and briefly describe the four fundamental components of a water system.

A. _____

_____

_____

_____

B. _____

_____

_____

_____

C. _____

_____

_____

_____

D. _____

_____

_____

_____

**9**

28. Identify the sizes of water mains for the following areas.

A. Residential _____

_____

_____

B. Business and industrial _____

_____

_____

29. Explain the operation of the following types of hydrants.

A. Dry barrel hydrant _____

_____

_____

_____

_____

_____

_____

B. Wet barrel hydrant _____

_____

_____

_____

_____

_____

_____

30. Identify the range of water flow for the various classes of hydrants by completing the following table.

| HYDRANT COLOR CODES | | |
|---|---|---|
| **Hydrant Class** | **Color** | **Flow** |
| Class AA | Light Blue | _____ |
| Class A | Green | _____ |
| Class B | Orange | _____ |
| Class C | Red | _____ |

**9**

31. List the three key elements to water shuttle operations.

    A. _____

    B. _____

    C. _____

32. List the four basic methods by which tankers unload water.

    A. _____

    B. _____

    C. _____

    D. _____

33. There are two important considerations regarding the establishment of a relay operation. List these two.

    A. _____

    _____

    _____

    B. _____

    _____

    _____

**SHORT ANSWER**

**Briefly answer each question in your own words.**

34. What are the fire department's main concerns regarding water treatment facilities?

    _____

    _____

    _____

    _____

35. Secondary feeders should be arranged in loops to give two directions of supply to any point. Why?

    _____

    _____

    _____

    _____

**9**

36. What is the function of a valve in a water distribution system?

_____

_____

37. What are two causes of increased friction loss in water mains?

_____

_____

38. What are the factors that have a major impact on the amount of flow of a hydrant?

_____

_____

39. When using a pitot tube and gauge to measure flow pressure, how far away from the orifice should the small opening or point of the pitot tube be held?

_____

_____

40. What is water shuttling, and when is it recommended?

_____

_____

_____

_____

_____

_____

*Courtesy of Tom McCarthy, Chicago, IL.*

# Fire Streams

# Fire Streams |10

## DEFINITIONS OF KEY TERMS

**Define each of the following terms.**

1. Fire stream

   _____

   _____

2. Friction loss

   _____

   _____

3. Water hammer

   _____

   _____

4. Velocity

   _____

   _____

5. Foam concentrate

   _____

   _____

6. Foam proportioner

   _____

   _____

7. Foam solution

   _____

   _____

8. Finished foam

   _____

   _____

**10**

9. Subsurface injection

_____

_____

---

**TRUE/FALSE**

**Mark each of the following statements true (T) or false (F). Correct each false statement.**

10. ☐ T ☐ F   Steam expansion is not rapid, but gradual.

_____

_____

11. ☐ T ☐ F   All fire streams must have a pressuring device, a hose, an agent, and a nozzle.

_____

_____

12. ☐ T ☐ F   A handline with a wide-angle fog pattern can be handled more easily than one with a straight-stream pattern.

_____

_____

13. ☐ T ☐ F   The piercing nozzle, chimney nozzle, and cellar nozzle are types of broken stream nozzles.

_____

_____

14. ☐ T ☐ F   Using a foam proportioner that is not matched to the foam maker has no significant effect on the foam produced.

_____

_____

15. ☐ T ☐ F   Foams designed for hydrocarbon fires will not extinguish polar solvent fires regardless of the concentration at which they are used.

_____

_____

16. ☐ T ☐ F   Fluoroprotein foam can be used to extinguish burning storage tanks by subsurface injection.

_____

_____

**10**

17. ☐ T ☐ F  Medium- and high-expansion foams have a high water content.

_____

_____

18. ☐ T ☐ F  The two basic pieces of equipment needed to produce a foam fire stream are a foam maker and a nozzle.

_____

_____

19. ☐ T ☐ F  The balanced pressure proportioner is the simplest and least expensive proportioning device.

_____

_____

20. ☐ T ☐ F  It is important for the proportioner and foam maker to match each other in order to produce usable foam.

_____

_____

## MULTIPLE CHOICE

**Circle the letter before the most appropriate response.**

21. Firefighters A and B are discussing broken streams. Firefighter A says that the droplets of a broken stream are larger than those of a fog stream. Firefighter B says that broken streams have greater penetration than fog streams and that broken streams can be useful where neither fog nor solid streams would be as effective. Who is correct?
    A. Firefighter A
    B. Firefighter B
    C. Both A and B
    D. Neither A nor B

22. Which of the following statements about fog stream nozzles is _not_ correct?
    A. An impinging-stream nozzle drives several jets of water together at a set angle to break the water into finely divided particles.
    B. A periphery-deflected stream is produced by deflecting water from the periphery of an inside, circular stem in a periphery-deflected fog nozzle.
    C. Fog stream nozzles are designed to be used for only one flow rate at a given discharge pressure.
    D. An impinging-stream nozzle usually produces a wide-angle fog pattern.

**10**

23. In general, multipurpose foams are designed to be used at a _____ percent concentration on hydrocarbon fires and at least a _____ percent concentration on polar solvent fires.
    A. 1½, 2, or 3; 1 to 6
    B. 1 to 3; 1 to 6
    C. 1 to 6; 6 to 10
    D. 6 to 10; 1 to 6

24. Medium-expansion foams are typically used at _____ percent concentrations.
    A. 1½, 2, or 3
    B. 3 to 10
    C. 1 to 6
    D. 6 to 10

25. Firefighters A and B are discussing the application of fluoroprotein foams. Firefighter A says that such foams must not be allowed to plunge into the fuel. Firefighter B says that fluoroprotein foams are ideally suited for using a plunge technique. Who is correct?
    A. Firefighter A
    B. Firefighter B
    C. Both A and B
    D. Neither A nor B

26. Firefighters A and B are discussing alcohol-resistant aqueous film forming foam (AFFF) and film forming fluoroprotein foam (FFFP). Firefighter A says that alcohol-resistant FFFP can be used on polar solvent fuels at 6 percent concentra tions and on hydrocarbon fuels at 3 percent. Firefighter B says that alcohol-resistant AFFF can be used on these fuels — 6 percent for polar solvents and 3 percent for hydrocarbons. Who is correct?
    A. Firefighter A
    B. Firefighter B
    C. Both A and B
    D. Neither A nor B

27. Which of the following types of foam can be used through nonaerating nozzles?
    A. Protein foams
    B. Fluoroprotein foams
    C. Medium-expansion foams
    D. AFFF

28. Which of the following types of foam is essentially a wetting agent that reduces the surface tension of water and allows it to soak into combustible materials easier than plain water?
    A. High-expansion foam
    B. Class A foam
    C. Hazardous materials vapor mitigating foam
    D. None of the above

**10**

29. The most effective appliance for generating low-expansion foam is the _____.
    A. Standard fixed-flow fog nozzle
    B. High-expansion foam generator
    C. Air aspirating foam nozzle
    D. Automatic nozzle

30. Firefighters A and B are discussing a foam fire stream system using an in-line eductor. Firefighter A says that if the eductor is attached directly to a pump discharge outlet, the ball valve gates should be closed. Firefighter B agrees and says that after the eductor suction hose is in place, the water supply pressure should be increased to 100 psi (700 kPa). Who is correct?
    A. Firefighter A
    B. Firefighter B
    C. Both A and B
    D. Neither A nor B

31. When preparing a foam fire stream system using an in-line eductor, make sure that the bottom of the concentrate is no more than _____ below the eductor.
    A. 1 foot (0.3 m)
    B. 6 inches (150 mm)
    C. 2 feet (0.6 m)
    D. 6 feet (2 m)

## IDENTIFICATION

**Identify the following items.**

32. Hydrocarbon fuels

_____

_____

33. Polar solvents

_____

_____

34. Master stream

_____

_____

35. Foam maker

_____

_____

**10** | Identify the characteristics of the different types of foam concentrates. Write the correct letters in the blanks; some blanks may have more than one correct answer.

36. _____ Has high heat resistance

37. _____ Can be freeze protected

38. _____ Can be made alcohol resistant

39. _____ Performance is not affected by freezing and thawing

40. _____ Can be stored premixed for a short period of time

41. _____ Has good low-temperature viscosity

42. _____ Is premixable in portable fire extinguishers and apparatus water tanks

43. _____ Can be stored at temperatures from 20°F to 120°F (-7°C to 49°C)

44. _____ Has good water retention

45. _____ Can be stored at temperatures from 10°F to 120°F (-12°C to 49°C)

46. _____ Is compatible with simultaneous application of dry chemical extinguishing agents

47. _____ Can be used with fresh or salt water

48. _____ Can be stored at temperatures from 25°F to 120°F (-4°C to 49°C)

A. Aqueous film forming foam (AFFF)

B. Film forming fluoroprotein foam (FFFP)

C. Fluoroprotein foam

D. Protein foam

**Provide the requested information.**

49. Identify the three major types of fire streams, and describe each.

A. _____

_____

B. _____

_____

C. _____

_____

50. Identify the nozzle pressure at which each of the following should operate.

A. Solid stream nozzle used on a handline _____

B. Solid stream master stream device _____

C. Fog nozzle _____

**10**

51. Identify the four methods by which foam extinguishes or prevents fire.

    A. _____

    _____

    B. _____

    _____

    C. _____

    _____

    D. _____

    _____

## LISTING

52. List the three observable results of the proper application of a fire stream.

    A. _____

    _____

    B. _____

    _____

    C. _____

    _____

53. List advantages and disadvantages of solid streams.

    _____

    _____

    _____

    _____

    _____

    _____

    _____

    _____

    _____

**10**

54. List advantages and disadvantages of fog streams.

_____
_____
_____
_____
_____
_____
_____
_____
_____
_____
_____

55. List the four elements that are necessary to produce high-quality fire fighting foam.
    A. _____
    B. _____
    C. _____
    D. _____

56. List characteristics of medium- and high-expansion foam concentrates.

_____
_____
_____
_____
_____
_____
_____
_____

57. List characteristics of Class A foams.

_____
_____
_____
_____
_____
_____
_____
_____
_____

**10**

58. List at least five of the most common reasons for generating poor-quality foam.

_____
_____
_____
_____
_____
_____
_____
_____
_____
_____
_____

## SHORT ANSWER

**Briefly answer each question in your own words.**

59. Why does the water of a fog stream absorb heat and convert into steam more rapidly than does the water of a straight stream?

_____
_____

60. How much does a cubic foot of water expand at 212°F (100°C)?

_____

61. How can water hammer be prevented?

_____

62. What two criteria are used to classify a stream as effective?

_____
_____
_____
_____
_____
_____

**10**

63. How does a straight stream differ from a solid stream?

_____

_____

_____

_____

64. What factors are important in producing a fog stream that will reach a desired objective?

_____

_____

_____

_____

65. What are the procedures for inspecting and maintaining nozzles?

_____

_____

_____

_____

_____

_____

_____

_____

_____

_____

66. There are two stages in the formation of foam. What takes place during each stage?

_____

_____

_____

_____

_____

_____

_____

67. What is hazardous materials vapor mitigating foam, and how does it work?

_____
_____
_____
_____
_____
_____

68. Why is it generally recommended that medium- and high-expansion foams not be used outdoors?

_____
_____

**10**

# 11

# Hose

# Hose | 11

## DEFINITIONS OF KEY TERMS

**Define each of the following terms.**

1. Fire hose

   _____

   _____

2. Shank

   _____

   _____

3. Hose bed

   _____

   _____

4. Dutchman

   _____

   _____

5. Hose load finish

   _____

   _____

## TRUE/FALSE

**Mark each of the following statements true (T) or false (F). Correct each false statement.**

6. ☐ T  ☐ F  The four basic types of construction of fire hose are braided, wrapped, woven jacket, and rubber covered.

   _____

   _____

7. ☐ T  ☐ F  If hose has not been unloaded from the apparatus during a period of 60 days, it should be removed, inspected, swept, and reloaded.

   _____

   _____

**11**

8. ☐ T ☐ F   Chemicals and chemical vapors often cause the hose lining and jacket to separate, but they will not damage the rubber lining.

_____

_____

9. ☐ T ☐ F   Most authorities feel that it is best for the inside of the rubber tube of the hose to be completely dry while in storage or on the fire apparatus.

_____

_____

10. ☐ T ☐ F   The male and female sides of a connected coupling can be distinguished from one another by noting the rocker lugs or pins on the shank — only female shanks have rocker lugs or pins.

_____

_____

11. ☐ T ☐ F   One way in which hose appliances differ from hose tools is that hose appliances have water flowing through them and tools do not.

_____

_____

12. ☐ T ☐ F   For most convenient handling, threaded-coupling hose must be arranged in the hose bed so that when hose is laid, the end with the female coupling is toward the water source, and the end with the male coupling is toward the fire.

_____

_____

13. ☐ T ☐ F   The front of the hose bed is that part of the compartment toward the rear of the apparatus, and the rear of the hose bed is that part of the compartment toward the front of the apparatus.

_____

_____

14. ☐ T ☐ F   For an accordion hose load, the first coupling placed in the hose bed should be located to the rear of the bed and can be put to either side if the bed is not split.

_____

_____

11

15. ☐ T ☐ F For a flat hose load, the load my be started on either side of a single bed or against the partition of a split hose bed.

_____

_____

16. ☐ T ☐ F Hose load finishes for forward lays are usually designed to speed the pulling of hose when making a hydrant connection and are not as elaborate as finishes for reverse lays.

_____

_____

17. ☐ T ☐ F The straight hose load finish is normally associated with a forward lay operation.

_____

_____

18. ☐ T ☐ F With the two-firefighter method of connecting couplings, the alignment of the hose must be done by the firefighter with the male coupling.

_____

_____

19. ☐ T ☐ F Making hydrant connections with a hard suction hose is considerably more difficult than with a soft intake hose.

_____

_____

20. ☐ T ☐ F When safely possible, hose should be advanced up stairways before it is charged with water.

_____

_____

21. ☐ T ☐ F The safest way to control a loose line is to attempt to put a kink in the hose some distance away from the loose end.

_____

_____

22. ☐ T ☐ F There is no need for a firefighter to stay at a master stream device when water is flowing.

_____

_____

**11**

23. ☐ T ☐ F  Acceptance testing should not be attempted by fire department personnel.

_____

_____

## MULTIPLE CHOICE

**Circle the letter before the most appropriate response.**

24. Fire hose is most commonly cut and coupled into lengths of _____ for convenience of handling and replacement but may be obtained in other lengths.
   A. 10 or 25 feet (3 m or 8 m)
   B. 25 or 50 feet (8 m or 15 m)
   C. 50 or 100 feet (15 m or 30 m)
   D. 100 or 200 feet (30 m or 60 m)

25. Which of the following types of hose is not subject to mold and mildew damage?
   A. Braided hose
   B. Rubber-jacket hose
   C. Woven-jacket hose
   D. Wrapped hose

26. Firefighters A and B are discussing chemical damage to fire hose. Firefighter A says that all chemicals can be removed by scrubbing the contaminated hose with a solution of baking soda and water. Firefighter B says that hose that has been exposed to hazardous materials and that cannot be decontaminated should be disposed of properly. Who is correct?
   A. Firefighter A
   B. Firefighter B
   C. Both A and B
   D. Neither A nor B

27. Which of the following statements about hose rolls is _not_ correct?
   A. The straight roll is commonly used for hose that is placed in storage.
   B. The donut roll is commonly used for hose that is going to be deployed for use directly from a roll.
   C. When a hose roll is completed, most commonly the female end of the hose is exposed.
   D. A hose roll completed so that the male end of the hose is exposed is often used when the hose will be reloaded on the apparatus for a reverse lay.

28. The primary advantage of using a _____ is that a pumper can remain at the incident scene so that its hose, equipment, and tools can be quickly obtained if needed.
   A. Split lay
   B. Reverse lay
   C. Forward lay
   D. Combination lay

29. Using a _____ is the most expedient way to lay hose if the apparatus that lays the hose must stay at the water source, as when boosting hydrant pressure to the supply line.
    A. Split lay
    B. Reverse lay
    C. Forward lay
    D. Combination lay

30. Which of the following statements about loading an accordion load into a split hose bed for a reverse lay is *not* correct?
    A. The first coupling placed in the hose bed should be located to the front of the bed.
    B. The first length of hose in the bed should be laid on edge against the partition.
    C. The female coupling of the first length of hose should be allowed to hang below the hose bed.
    D. The folds should be staggered so that the end of every other fold is approximately 2 inches (50 mm) from the edge of the bed.

31. Which of the following statements about the horseshoe hose load is correct?
    A. The horseshoe load is recommended for large diameter hose.
    B. One person can easily make shoulder folds for a hose carry from a horseshoe load.
    C. With a horseshoe load, the hose sometimes comes out in a wavy or snakelike lay in the street as it is pulled from the hose bed.
    D. The primary disadvantage of the horseshoe load is that it has more sharp bends than either the accordion or the flat load.

32. Firefighters A and B are discussing the horseshoe hose load. Firefighter A says that in a single hose bed, the load may be started on either side. Firefighter B says that in a split hose bed, the first length should be laid against the partition rather than the side. Who is correct?
    A. Firefighter A
    B. Firefighter B
    C. Both A and B
    D. Neither A nor B

33. Which of the following hose loads is easiest to load and is also suited for any size of supply hose?
    A. Accordion load
    B. Horseshoe load
    C. Flat load
    D. Split bed load

34. Firefighters A and B are discussing the flat hose load. Firefighter A says that for a forward lay, the first length of hose should be laid flat in the bed with the female coupling hanging below the hose bed. Firefighter B says that the folds of the second tier should be approximately 2 inches (50 mm) shorter than the folds of the first tier and that the third tier should be even with the first. Who is correct?
    A. Firefighter A
    B. Firefighter B
    C. Both A and B
    D. Neither A nor B

**11**

35. Firefighters A and B are discussing preconnected hoselines. Firefighter A says that preconnected attack lines may be carried in tailboard compartments, side compartments, or reels. Firefighter B says that preconnected hoselines are generally 50 feet (15 m) in length. Who is correct?
    A. Firefighter A
    B. Firefighter B
    C. Both A and B
    D. Neither A nor B

36. Which of the following is a nonpreconnected hose load that allows the firefighter to load hose from the hose bed directly onto the shoulder?
    A. Horseshoe load
    B. Minuteman load
    C. Accordion load
    D. None of the above

37. When advancing a dry hoseline up a stairway, where should the firefighter lay the hose?
    A. Along the hand rail on the inside wall
    B. On the stairs midway between the walls
    C. On the stairs against the inside wall
    D. On the stairs against the outside wall

38. Firefighters A and B are discussing the advancement of hose up a stairway. Firefighter A says that hose should be advanced before it is charged if it is safely possible. Firefighter B says that to advance a charged line, the firefighter should first clamp the line. Who is correct?
    A. Firefighter A
    B. Firefighter B
    C. Both A and B
    D. Neither A nor B

39. Which of the following statements concerning the advancement of a hoseline down a stairway is *not* correct?
    A. It is easier to advance a dry hoseline than to advance a charged hoseline.
    B. A charged hoseline should be advanced in most cases.
    C. It is recommended that firefighters be stationed only at the nozzle and at the top of the stairway.
    D. Advancing an uncharged line downstairs is recommended only when it is evident that no fire — or that only a very minor fire — is present.

40. To extend a charged hoseline, the firefighter must first _____.
    A. Crack the nozzle open slightly and apply a hose clamp behind the nozzle
    B. Close the nozzle and apply a hose clamp behind the nozzle
    C. Remove the nozzle
    D. Have the pump operator slightly reduce the pressure on the hoseline

41. When should fire department hose be service tested?
    A. Annually
    B. After being repaired
    C. After being run over by a vehicle
    D. All of the above
42. During service testing of fire department hose, the test pressure should be maintained for _____ before reducing the pump pressure and closing the discharge valves.
    A. 5 minutes
    B. 10 minutes
    C. 30 minutes
    D. 1 hour

## IDENTIFICATION

**Identify the following items.**
43. NFPA 1961

_____

_____

44. NFPA 1963

_____

_____

45. NFPA 1901

_____

_____

46. NFPA 1962

_____

_____

47. Higbee cut

_____

_____

48. Higbee indicator

_____

_____

49. Storz coupling

_____

_____

**11**

50. Accordion load

_____
_____

51. Horseshoe load

_____
_____

52. Flat load

_____
_____

53. Straight finish

_____
_____

54. Reverse horseshoe finish

_____
_____

**Identify the following hose tools and appliances, and identify the use of each.**

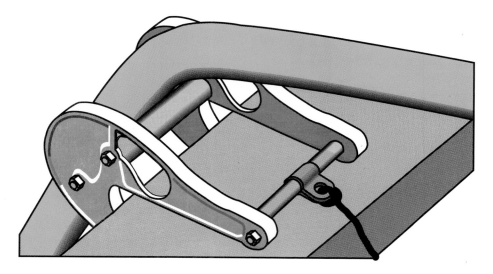

55. A. Name

    B. Use _____

_____
_____
_____

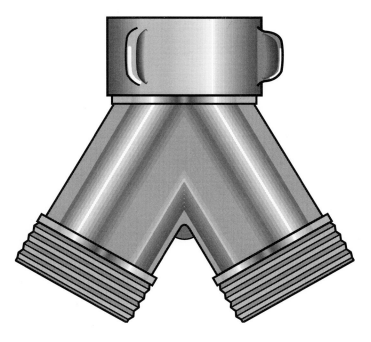

56. A. Name
    B. Use _____

_____

_____

_____

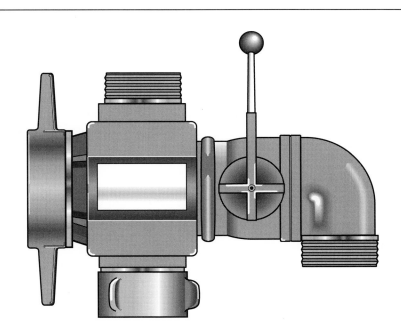

57. A. Name
    B. Use _____

_____

_____

_____

**11**

58. A. Name
    B. Use _____
    _____
    _____
    _____

59. A. Name
    B. Use _____
    _____
    _____
    _____

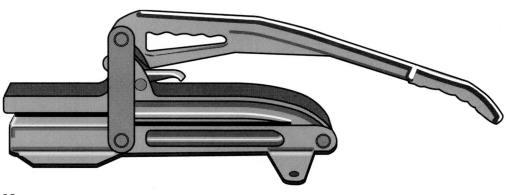

60. A. Name _____

    B. Use _____

    _____

    _____

    _____

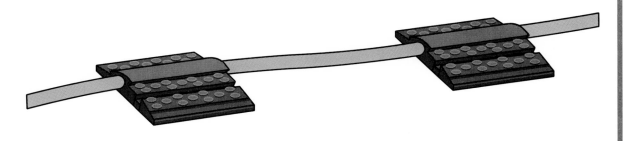

61. A. Name _____

    B. Use _____

    _____

    _____

    _____

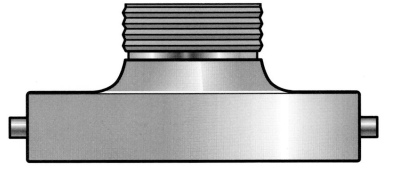

62. A. Name _____

    B. Use _____

    _____

    _____

    _____

**11**

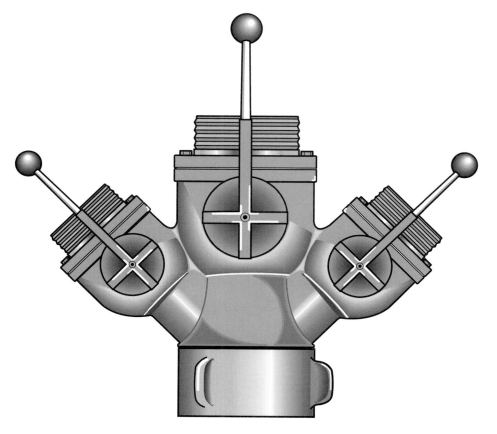

63. A. Name _____

    B. Use _____

    _____

    _____

    _____

    _____

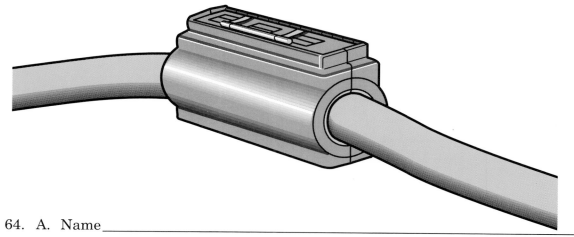

64. A. Name _____

    B. Use _____

    _____

    _____

    _____

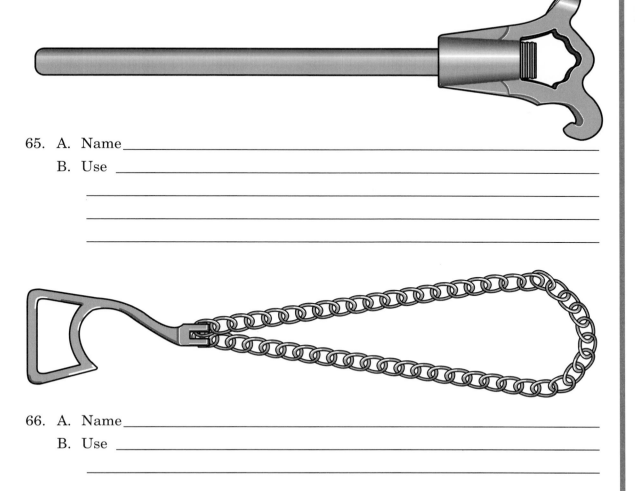

65. A. Name _____
    B. Use _____
    _____
    _____
    _____

66. A. Name _____
    B. Use _____
    _____
    _____
    _____

**Provide the requested information.**

67. Identify the sizes and lengths of hose that NFPA 1901 requires pumpers to carry.

_____
_____
_____
_____
_____
_____
_____
_____
_____
_____

**11**

68. Identify the three basic hose lays for supply hose, briefly describe each, and tell when each is used.

    A. _____
    _____
    _____
    _____
    _____

    B. _____
    _____
    _____
    _____
    _____

    C. _____
    _____
    _____
    _____
    _____

69. Identify the tools and appliances needed by the person "catching" the hydrant.

    _____
    _____

## LISTING

70. List at least five ways of preventing mechanical damage to fire hose.

    _____
    _____
    _____
    _____
    _____
    _____
    _____
    _____
    _____
    _____
    _____

**11**

71. List at least five ways of preventing thermal damage to fire hose.

_____

_____

_____

_____

_____

_____

_____

_____

72. List the three ways in which woven-jacket hose may be dried after it has been thoroughly washed.

A. _____

B. _____

C. _____

73. List general guidelines to follow when loading hose.

_____

_____

_____

_____

_____

_____

_____

_____

_____

_____

_____

_____

_____

_____

_____

_____

_____

_____

**11**

74. List the general safety guidelines that firefighters should observe when advancing a hoseline into a burning structure.

_____
_____
_____
_____
_____
_____
_____
_____

75. List the three main uses for a master stream.
    A. _____
    B. _____
    C. _____

## SHORT ANSWER

**Briefly answer each question in your own words.**

76. What equipment may be used to wash fire hose?

_____
_____
_____
_____

77. How are hose couplings cleaned?

_____
_____
_____
_____
_____
_____
_____
_____

78. What are five-piece couplings, and when are they used?

_____
_____
_____
_____

**11**

79. What is the procedure for using the foot-tilt method to connect fire hose couplings?

_____
_____
_____
_____
_____
_____
_____
_____

80. By using either of what two methods can firefighters break a tight coupling when spanner wrenches are not available?

_____
_____

81. The preconnected flat load is similar to the flat load for larger supply hose with two exceptions. What are these exceptions?

_____
_____
_____
_____

82. What is the major disadvantage of the triple layer load?

_____
_____
_____
_____

83. What is the primary advantage of the minuteman load?

_____
_____
_____
_____

84. The firefighter who is to make the hydrant connection must be knowledgeable in what two things?

_____
_____

85. When making a hydrant connection from a forward lay, at what point can the firefighter catching the hydrant motion the driver/operator to proceed?

_____
_____

**11**

86. What is the advantage of using a four-way hydrant valve?

_____

_____

_____

_____

_____

87. How does the firefighter pull the preconnected flat hose load?

_____

_____

_____

_____

_____

_____

_____

_____

88. How does the firefighter pull the minuteman load?

_____

_____

_____

_____

_____

_____

_____

_____

_____

89. When bringing hose up a stairway to make a standpipe connection, where should firefighters stop to make the connection?

_____

_____

90. How should firefighters advance a charged hoseline up a ladder?

_____

_____

_____

_____

_____

**11**

91. After a hoseline has been under pressure, two sections of hose should be used to replace one bad section of hose. Why?

_____

_____

_____

_____

_____

92. How should a single firefighter position herself or himself in order to properly handle a medium-sized hose and nozzle?

_____

_____

_____

_____

_____

_____

_____

_____

_____

93. When two firefighters are handling a nozzle on a medium-size attack line, how should the backup firefighter be positioned?

_____

_____

_____

_____

_____

94. What is the purpose of service testing fire department hose?

_____

_____

_____

_____

Courtesy of Ron Jeffers, Union City, NJ.

# Fire Control

# Fire Control |12

**Define each of the following terms.**

1. Flammable liquids

   _____

   _____

2. Combustible liquids

   _____

   _____

3. BLEVE

   _____

   _____

4. Ground radient

   _____

   _____

5. Topography

   _____

   _____

6. Slope aspect

   _____

   _____

**Mark each of the following statements true (T) or false (F). Correct each false statement.**

7. ☐ T ☐ F Whenever operating in a hazardous or potentially hazardous location on the emergency scene, firefighters should work in pairs.

   _____

   _____

**12**

8. ☐ T ☐ F   At a structural fire, firefighters should always approach and attack the fire from the burned side to keep reignition to a minimum.

_____

_____

9. ☐ T ☐ F   In most incidents, the hose crew backing out of an area of a structure should keep the stream operating until all personnel are out.

_____

_____

10. ☐ T ☐ F   An indirect attack is not desirable where victims may remain trapped or where the spread of fire to uninvolved areas cannot be contained.

_____

_____

11. ☐ T ☐ F   Hydrocarbon liquids will not mix with water, but polar solvents will.

_____

_____

12. ☐ T ☐ F   When approaching a flammable liquid or gas storage vessel after its relief valve closes, firefighters should approach from the ends of the vessel and not at right angles to it.

_____

_____

13. ☐ T ☐ F   All LPG containers are subject to BLEVEs when exposed to intense heat or open flame.

_____

_____

14. ☐ T ☐ F   Unburned gas may be dissipated by a straight stream of at least 150 gpm (600 L/min).

_____

_____

15. ☐ T ☐ F   Firefighters should not enter a manhole except to attempt a rescue or to fight a fire.

_____

_____

**12**

16. ☐ T ☐ F  Firefighters should treat all electrical wires as though the wires are energized and of high voltage.

_____

_____

17. ☐ T ☐ F  Water is never effective in suppressing combustible metal fires.

_____

_____

18. ☐ T ☐ F  With a basement fire, the ideal ventilation point is at a point near the stairs that the firefighters will descend.

_____

_____

19. ☐ T ☐ F  A lifeline should be tied to each rescuer entering a confined enclosure, and this line must be monitored constantly.

_____

_____

20. ☐ T ☐ F  Regarding wildland fires, the size of the fuel will determine the fire's intensity and the amount of water needed for extinguishment.

_____

_____

## MULTIPLE CHOICE

**Circle the letter before the most appropriate response.**

21. Before entering the fire area, the person at the nozzle should _____.
    A. Bleed the air from the line by opening the nozzle slightly
    B. Check operation of the nozzle
    C. Set a proper pattern for the attack
    D. All of the above

22. Which of the following statements about attacking structural fires is correct?
    A. In an area not yet ventilated, it is important to use a 30-degree fog pattern.
    B. When ventilation ahead of the nozzle is provided, a fog stream should _not_ be used.
    C. In an unventilated setting, a straight stream does not upset the thermal layering as much as a fog stream does.
    D. None of the above are correct.

**12**

23. Firefighters A and B are discussing the advancement of the hoseline team. Firefighter A says that the nozzle should always be opened as soon as the firefighter at the nozzle enters the structure. Firefighter B disagrees and says that the nozzle should *not* be opened until fire is encountered. Who is correct?
    A. Firefighter A
    B. Firefighter B
    C. Both A and B
    D. Neither A nor B

24. The most efficient use of water on free-burning fires is made by _____.
    A. A combination attack
    B. A direct attack
    C. An indirect attack
    D. A fog pattern attack

25. IFSTA recommends that interior fires and larger exterior fires, including automobile fires, be attacked with no less than a _____ hoseline.
    A. 1-inch (25 mm)
    B. 1½-inch (38 mm)
    C. 2½-inch (65 mm)
    D. 3-inch (77 mm)

26. When using water to "move" flammable liquid fire, the firefighters should _____.
    A. Plunge the fire stream into the liquid
    B. "Sweep" the fuel by quickly directing the nozzle from side to side
    C. Keep the leading edge of the fog pattern in contact with the fuel surface
    D. Use a solid stream

27. Firefighters A and B are discussing incidents involving LPG and natural gas. Firefighter A says that natural gas and LPG have the same characteristics, including both being heavier than air, and therefore firefighters can easily handle such an incident without outside help. Firefighter B says that the local utility should be contacted when any emergency involving natural gas occurs in its service area. Who is correct?
    A. Firefighter A
    B. Firefighter B
    C. Both A and B
    D. Neither A nor B

28. Which of the following statements about an incident involving a pipe break in a natural gas distribution system is *not* correct?
    A. One of the firefighter's first concerns should be the evacuation of the area around the break and downwind from it.
    B. If the gas is burning, firefighters should extinguish the flame.
    C. Firefighters should not attempt to operate main valves.
    D. Firefighters may stop the flow of gas into a building by turning the cutoff valve to the closed position.

**12**

29. Which of the following is a possible consequence of electrical shock?
    A. Cardiac arrest
    B. Respiratory arrest
    C. Damage to joints
    D. All of the above

30. At fires involving electrical transformers and broken transmission lines, firefighters should _____.
    A. Allow transformers aboveground to burn themselves out
    B. Clear an area equal to one-half the height of a pole
    C. Carefully extinguish a burning transformer at ground level with a dry chemical extinguisher
    D. None of the above

31. During structural fire fighting operations, it is desirable for electrical power _____.
    A. To remain on until all firefighters are out of the building
    B. To be shut off as soon as fire personnel arrive at the scene
    C. To remain on until the building becomes damaged to the point that service is interrupted or an electrical hazard exists
    D. To be shut off unless needed to provide lighting

32. Firefighters A and B are discussing electrical emergencies. Firefighter A says that a firefighter must not touch any vehicle or apparatus that is in contact with electrical wires. Firefighter B agrees and says that if a firefighter must leave an apparatus that is already charged, he or she should jump clear of the apparatus, never touching it and the ground at the same time. Who is correct?
    A. Firefighter A
    B. Firefighter B
    C. Both A and B
    D. Neither A nor B

33. The _____ will usually initiate incident command and the fire attack.
    A. First-due engine company
    B. Second-due engine company
    C. Truck company
    D. Battalion/district chief

34. The _____ is responsible for forcible entry, search and rescue, ventilation, and securing utilities.
    A. First-due engine company
    B. Second-due engine company
    C. Truck company
    D. Squad/rescue company

**12**

35. Unless otherwise assigned, the first responsibility of the _____ is to make sure that an adequate water supply to the fireground is established.
    A. First-due engine company
    B. Second-due engine company
    C. Truck company
    D. Water supply officer

36. It is generally desirable to have the _____ personnel assigned as interior or exterior team members — the interior team to perform search and rescue and the exterior team to raise ladders to enter or ventilate the building from the outside.
    A. First-due engine company
    B. Second-due engine company
    C. Truck company
    D. Squad/rescue company

37. When fighting a fire in a multistory structure, the fire attack typically will be initated from _____ the fire floor, and the staging of extra equipment and personnel will usually be established _____.
    A. One floor below; one floor below
    B. One floor below; two floors below
    C. Two floors below; two floors below
    D. One floor below; one floor above

38. Which of the following statements about confined enclosure incidents is *not* correct?
    A. The staging area should be near the entrance.
    B. A safety officer should be stationed at the entrance to keep track of personnel entering and leaving the enclosure.
    C. Rescuers must be able to use the selected communication system without removing their SCBA masks.
    D. Rescuers should wear SCBA only if the atmosphere is hazardous.

39. Firefighters A and B are discussing how topography affects wildland fires. Firefighter A says that fires will usually move faster downhill than uphill. Firefighter B agrees and says that the steeper the slope, the slower the fire will move. Who is correct?
    A. Firefighter A
    B. Firefighter B
    C. Both A and B
    D. Neither A nor B

40. Which part of a wildland fire travels or spreads most rapidly?
    A. Flank
    B. Finger
    C. Head
    D. Perimeter

**12**

41. Firefighters A and B are discussing attacking wildland fires. Firefighter A says that the control line may be established at the burning edge of the fire, next to it, or at a considerable distance away. Firefighter B says the objective is to establish fire breaks that completely encircle the fire. Who is correct?
   A. Firefighter A
   B. Firefighter B
   C. Both A and B
   D. Neither A nor B

## IDENTIFICATION

**Identify the following items.**

42. Company officer

_____

_____

43. Apparatus driver/operator

_____

_____

44. "O" pattern

_____

_____

45. LPG

_____

_____

46. SOPs

_____

_____

47. O-A-T-H method

_____

_____

48. Perimeter

_____

_____

# 12

**Identify the following characteristics as those of LPG, natural gas, or both.**

49. _____ Has no natural odor

50. _____ Is lighter than air

51. _____ Is nontoxic

52. _____ Has a distinctive odor added to it

53. _____ Is explosive in concentrations between 4 percent and 14 percent

54. _____ Is classified as an asphyxiant

55. _____ Is distributed from gas wells to its points of usage by a nationwide network of surface and subsurface pipes

56. _____ Is about one and one-half times as heavy as air

57. _____ Is explosive in concentrations between 1.5 percent and 10 percent

58. _____ Is stored in tanks or cylinders near its point of usage and flows from its storage to the appliances it serves through underground piping and copper tubing

59. _____ Is used primarily as a fuel gas in campers, mobile homes, and rural homes

**Identify the parts of the wildland fire illustrated below. Write the correct names in the blanks.**

| | |
|---|---|
| Left flank | Right flank |
| Finger | Spot fire |
| Head | Rear |

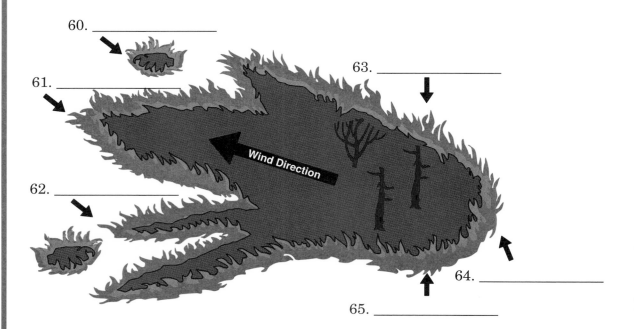

60. _____

61. _____

62. _____

63. _____

64. _____

65. _____

Wind Direction

**Provide the requested information.**

66. Describe the following three methods of water application for structural or Class A fires.

    A. Direct attack _____

    _____
    _____
    _____
    _____

    B. Indirect attack _____

    _____
    _____
    _____
    _____

    C. Combination attack _____

    _____
    _____
    _____
    _____

67. Describe the two ways in which firefighters can extinguish basement fires in instances where the basement is inaccessible.

    A. _____
    _____

    B. _____
    _____

68. List examples of the following types of ground cover fuels.

    A. Ground fuels (duff)_____

    _____
    _____

    B. Surface fuels _____

    _____
    _____

    C. Crown fuels _____

    _____
    _____

**12**

# 12 | LISTING

69. List potential hazards of which structural attack team members must beware.

_____

_____

_____

_____

_____

_____

_____

_____

_____

_____

_____

_____

70. List factors to consider when selecting the correct hoseline for attacking a structural fire.

_____

_____

_____

_____

_____

_____

_____

_____

71. List the four ways in which water can be used when suppressing Class B fires.

A. _____

_____

B. _____

_____

C. _____

_____

D. _____

_____

**12**

72. The handling of fires in flammable liquid storage facilities differs from handling those in vehicles transporting flammable fuels. List at least five of the major differences.

_____

_____

_____

_____

_____

_____

73. List the six factors most affecting the seriousness of electrical shock.

A. _____

_____

B. _____

_____

C. _____

_____

D. _____

_____

E. _____

_____

F. _____

_____

74. SOPs should be established to follow the most commonly accepted order of fireground priorities. List in order these three priorities.

A. _____

B. _____

C. _____

75. The officer of the first company will position the initial attack line so that certain priorities are covered or considered. List these priorities.

_____

_____

_____

_____

_____

_____

**12**

76. List the priorities of the second-due engine company after it has seen that an adequate water supply has been established.

_____

_____

_____

_____

_____

77. List the three most important factors that affect wildland fire behavior.

A. _____

B. _____

C. _____

## SHORT ANSWER

**Briefly answer each question in your own words.**

78. In addition to operating fire streams, teams advancing hoselines should also carry equipment necessary to force entry or perform other tasks. At a minimum, what basic equipment should be carried?

_____

_____

_____

_____

79. With what types of fires are booster lines used?

_____

_____

80. Besides the possibility of the fuel igniting, why should firefighters avoid standing in pools of flammable or combustible liquids?

_____

_____

_____

_____

_____

**12**

81. What are the appropriate actions for firefighters to take if a leaking flammable liquid is burning around relief valves or piping?

_____

_____

_____

_____

_____

_____

_____

82. When are BLEVEs most likely to occur?

_____

_____

_____

83. From what sources may firefighters gain information regarding the nature of the cargo of a transport vehicle?

_____

_____

_____

_____

84. Should firefighters extinguish burning gas that is leaking from a gas line? Why or why not?

_____

_____

_____

_____

85. At the scene of an electrical fire, what action should be taken before initiating fire suppression activities?

_____

_____

86. Why should water not be used — even in the form of fog — to extinguish fires involving high-voltage installations?

_____

_____

**12**

87. What are the major responsibilities of the chief officer?

_____

_____

_____

_____

_____

_____

88. What is the basic procedure for attacking a vehicle fire?

_____

_____

_____

_____

89. What hazard do spot fires present to personnel and equipment working on the main fire?

_____

_____

_____

## CASE STUDIES

*Case Study 1*

Fire Dispatch receives a report of a dwelling fire at 620 N.W. 34th Street. Engines 5 and 11, Truck 5, Rescue Squad 18, and District Chief (DC) 602 are dispatched. Each company rides with four personnel, and DC 602 has a driver. First-in Engine 5 notices a loom-up of smoke as it approaches the scene. They stop at a hydrant, wrap the hydrant with the hose and Humat valve, and lay into the scene. Once on the scene, Engine Captain 5 reports a one-story frame dwelling with fire visible in the left rear of the dwelling. Upon exiting the cab, the captain is advised by a distraught woman that the fire is in the left rear bedroom and that her five-year-old son is in the right rear bedroom. Other companies will arrive in the following order: DC 602, Engine 11 and Truck 5 (simultaneously), and Rescue Squad 18.

Based on this scenario and the information contained in the chapter, what actions should each of the companies take from this point on? (**NOTE:** The chapter does not contain information on Rescue Squad procedures, so you are free to use them as desired.)

**12**

A.   Engine 5 _____
_____
_____
_____
_____
_____
_____
_____
_____

B.   DC 602 _____
_____
_____
_____
_____
_____
_____
_____
_____
_____
_____
_____

C.   Engine 11 _____
_____
_____
_____
_____
_____
_____
_____
_____
_____
_____

**12**

D. Truck 5 _____

_____

_____

_____

_____

_____

_____

_____

_____

_____

_____

_____

E. Rescue Squad 18 _____

_____

_____

_____

_____

_____

_____

_____

_____

_____

_____

_____

_____

_____

*Case Study 2*
    Engine and Tanker 33 are dispatched to a reported car fire in a rural area of their district. (**NOTE:** Tanker 33 has a 500 gpm [2 000 L/min] pump and 3,000 gallons [12 000 L] of water). Temperatures are moderate for an autumn afternoon, although a 15- to 20-mile-per-hour (24 km/h to 32 km/h) wind is blowing. While Engine and Tanker 33 are en route, dispatch receives a second call reporting that a brush fire is now starting. Dispatch adds Brush Pumpers 33 and 20 to the assignment. Figure 12-A shows the conditions found when Engine and Tanker 33 arrive. About one acre of the field is on fire.

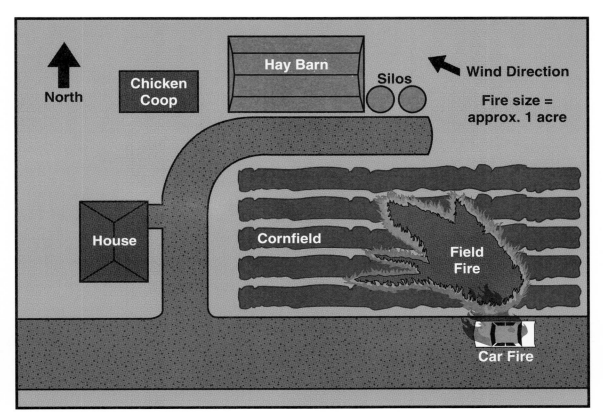

**Figure 12-A**

A.   The scene shows two fires in progress. Which must be given priority? Based on the time of the year, what must be a concern to the firefighters?

_____

_____

_____

_____

_____

_____

_____

_____

_____

**12**

B.  Give a suggested plan of action for Engine and Tanker 33, illustrating the plan on Figure 12-A.

_____

_____

_____

_____

_____

_____

_____

_____

_____

C.  What will Brush Pumpers 33 and 20 do upon arrival? Would a direct or indirect attack be more appropriate?

_____

_____

_____

_____

_____

_____

_____

_____

_____

_____

_____

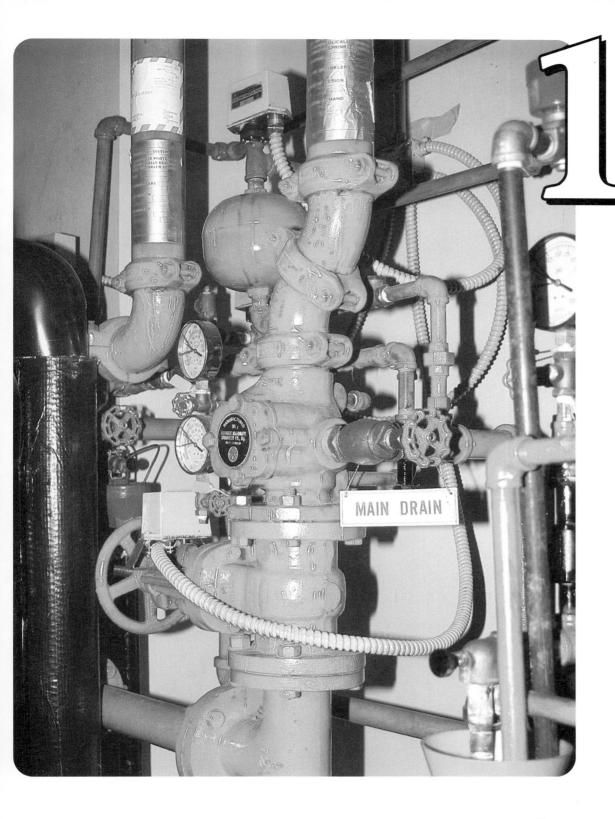

MAIN DRAIN

# 13

# Automatic
# Sprinkler
# Systems

# Automatic Sprinkler Systems | 13

## DEFINITIONS OF KEY TERMS

**Define each of the following terms.**

1. Indicating valve

   _____

   _____

2. Accelerator

   _____

   _____

3. Exhauster

   _____

   _____

4. Retarding device

   _____

   _____

## TRUE/FALSE

**Mark each of the following statements true (T) or false (F). Correct each false statement.**

5. ☐ T   ☐ F   The life safety of building occupants is enhanced by the presence of a sprinkler system because it discharges water directly on the fire while the fire is usually in the incipient stage.

   _____

   _____

6. ☐ T   ☐ F   Decorative sprinklers, such as plated or ceiling sprinklers, are not required to be color coded.

   _____

   _____

**13**

7. ☐ T  ☐ F  The three basic sprinkler designs are upright, pendant, and sidewall; any of which may be replaced with one of another design.

_____

_____

8. ☐ T  ☐ F  The main control valve of the sprinkler system should always be returned to the closed position after maintenance is complete.

_____

_____

9. ☐ T  ☐ F  Main water control valves are nonindicating valves that are automatically operated.

_____

_____

10. ☐ T  ☐ F  The water supply for sprinkler systems is designed to supply all of the sprinklers actually installed on the system.

_____

_____

11. ☐ T  ☐ F  Residential sprinkler systems are installed in one- and two-family dwellings and may be either wet or dry systems.

_____

_____

## MULTIPLE CHOICE

**Circle the letter before the most appropriate response.**

12. The _____ is the vertical piping to which the sprinkler valve, one-way check valve, fire department connection, alarm valve, main drain, and other sprinkler system components are attached.
    A.  Feed main
    B.  Distributor
    C.  Riser
    D.  Cross main

13. Fire department sprinkler connections should be supplied with water from pumpers that have a capacity of at least _____ or greater.
    A.  500 gpm (2 000 L/min)
    B.  1,000 gpm (4 000 L/min)
    C.  1,250 gpm (5 000 L/min)
    D.  1,500 gpm (6 000 L/min)

14. A _____ system is the simplest type of automatic fire sprinkler system and generally requires little maintenance.
    A. Deluge
    B. Wet-pipe
    C. Dry-pipe
    D. Residential

15. Which of the following sprinkler systems has sprinklers that do not have heat-responsive elements?
    A. Deluge
    B. Wet-pipe
    C. Dry-pipe
    D. Residential

16. When fighting fires in occupancies that have operating sprinkler systems, firefighters should _____.
    A. Make sure that all sprinkler heads are discharging water
    B. Check control valves to see that they are open
    C. Ensure that pressure at the fire department connection is in the range of 150 psi (1 050 kPa)
    D. All of the above

17. Sprinkler control valves should not be closed until _____.
    A. Pumpers have been disconnected
    B. A hoseline team is in position to attack the fire
    C. It is determined that the fire has been extinguished
    D. All of the above

18. Preferably, sprinkler equipment should be restored to service before fire department personnel leave the premises. Who should perform sprinkler system maintenance?
    A. Fire department personnel
    B. Sprinkler manufacturer representative
    C. Occupant representative
    D. None of the above

## IDENTIFICATION

**Identify the following items.**
19. NFPA 13

_____

_____

20. NFPA 13D

_____

_____

**13**

21. OS&Y valve

_____

_____

22. PIV

_____

_____

23. WPIV

_____

_____

24. PIVA

_____

_____

25. Fire department connection

_____

_____

26. Ball drip valve

_____

_____

**Identify the functions of the following sprinkler system components.**

27. Sprinklers _____

_____

_____

28. Control valves _____

_____

_____

29. Alarm test valve _____

_____

_____

30. Inspector's test valve _____

_____

_____

31. Waterflow alarms _____
_____
_____

**Identify the use and operation of each of the following types of sprinkler systems.**
32. Wet-pipe system
    A. Use _____
    B. Operation _____
_____
_____
_____
_____
_____
_____
_____

33. Dry-pipe system
    A. Use _____
    B. Operation _____
_____
_____
_____
_____
_____
_____
_____

34. Pre-action system
    A. Use _____
    B. Operation _____
_____
_____
_____
_____
_____
_____
_____

**13**

35. Deluge system
   A. Use _____
   B. Operation _____
   _____
   _____
   _____
   _____
   _____
   _____
   _____

**Identify the components of the following wet-pipe sprinkler system. Write the correct names in the blanks.**

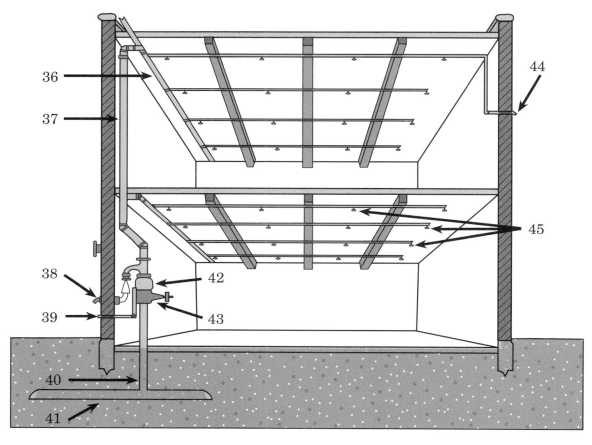

36. _____    41. _____
37. _____    42. _____
38. _____    43. _____
39. _____    44. _____
40. _____    45. _____

## LISTING

46. List reasons why a sprinkler system may not perform properly.

_____

_____

_____

_____

_____

_____

47. List the five common types of water supplies used as primary and secondary water supplies for sprinkler systems.

A. _____

B. _____

C. _____

D. _____

E. _____

## SHORT ANSWER

**Briefly answer each question in your own words.**

48. What is the difference between the coverage provided by a complete sprinkler system and that provided by a partial sprinkler system?

_____

_____

_____

_____

49. The selection of sprinklers for a given application should be based upon what two factors?

_____

_____

_____

_____

50. How are sprinklers activated to release water?

_____

_____

_____

**13**

51. Where is the sprinkler system control valve usually located?

_____

_____

_____

_____

52. With an OS&Y valve, is the threaded stem out of or inside the yoke when the valve is open?

_____

53. What is the function of the fire department connection check valve?

_____

_____

54. How is the proper direction of water flow through a check valve denoted?

_____

_____

55. What does it mean to the firefighter if the air pressure gauge and the water pressure gauge of a dry-pipe system read the same?

_____

_____

56. What is the danger of prematurely closing the sprinkler system control valve? What precaution should be taken when it is closed?

_____

_____

57. Firefighters may find it necessary to stop the flow of water from a single sprinkler that has been activated. Without closing the main water control valve, how can the firefighter do this?

_____

_____

_____

_____

# Salvage And Overhaul

# Salvage And Overhaul |14

## DEFINITIONS OF KEY TERMS

**Define each of the following terms.**

1. Salvage

   _____

   _____

2. Overhaul

   _____

   _____

## TRUE/FALSE

**Mark each of the following statements true (T) or false (F). Correct each false statement.**

3. ☐ T ☐ F  It is common for truck company personnel to perform salvage operations; however, engine company personnel may do salvage work in some jurisdictions.

   _____

   _____

4. ☐ T ☐ F  Only one firefighter is needed to fold a salvage cover for a one-firefighter spread.

   _____

   _____

5. ☐ T ☐ F  The balloon throw is the most common method for two firefighters to deploy a salvage cover.

   _____

   _____

6. ☐ T ☐ F  By using it as a catchall, a salvage cover may be used to catch and route water away from fire fighting operations.

   _____

   _____

**14**

7. ☐ T ☐ F  Effective water chutes can be made with two pike poles and a salvage cover.

_____

_____

8. ☐ T ☐ F  To open a plaster ceiling from below, the firefighter, using either a pike pole or plaster hook, must first break the plaster and then pull off the lath.

_____

_____

9. ☐ T ☐ F  Typically, at least 1½-inch (38 mm) or larger attack lines are used for overhaul.

_____

_____

## MULTIPLE CHOICE

**Circle the letter before the most appropriate response.**

10. Which of the following statements about salvage and overhaul is correct?
    A. One of the most effective means of building goodwill within the community is performing proper salvage and overhaul.
    B. Salvage operations consist of procedures used to locate hidden fires.
    C. Salvage and overhaul at times may be performed simultaneously, but salvage operations generally follow those of overhaul.
    D. None of the above are correct.

11. At what time should fire department personnel begin salvage operations?
    A. Immediately after the fire has been extinguished
    B. As soon as possible after the fire is under control
    C. Before or at the same time the fire attack is initiated
    D. As soon as overhaul operations have been completed

12. Firefighters A and B are discussing salvage operations. Firefighter A says that salvage operations must never take priority over fire attack. Firefighter B says that in instances where the value of the building contents may exceed the replacement cost of the burning section of the structure, it may be prudent to delay the fire attack until some salvage work can be done. Who is correct?
    A. Firefighter A
    B. Firefighter B
    C. Both A and B
    D. Neither A nor B

**14**

13. Firefighters A and B are discussing water removal from a structure. Firefighter A says that the best way to remove water from basements and elevator shafts is to use the fire department pumpers. Firefighter B says that portable water pumps should be used to remove water from these areas. Who is correct?
    A. Firefighter A
    B. Firefighter B
    C. Both A and B
    D. Neither A nor B

14. Which of the following statements about salvage operations is *not* correct?
    A. Salvage operations include covering doors that have been broken.
    B. Firefighters should cover openings of the building to prevent further damage to the property by weather.
    C. Firefighters should make no attempt to repair or cover broken windows but should advise the occupant or owner to do so as soon as the scene has been secured.
    D. Roofing or tar paper can be nailed down using nails with laths to cover roof openings.

15. A firefighter opening a wall during overhaul would most likely use which of the following tools?
    A. Axe
    B. Hook
    C. Pike pole
    D. Plaster hook

16. To open a ceiling to check for fire extension, which of the following would be the appropriate tool for the firefighter to use?
    A. Axe
    B. Hook
    C. Pike pole
    D. None of the above

17. Firefighters A and B are discussing overhaul operations. Firefighter A says that it is important to check insulation materials because they can harbor hidden fires for a prolonged period of time. Firefighter B says that it is usually necessary to remove these materials to check them properly or extinguish any fire. Who is correct?
    A. Firefighter A
    B. Firefighter B
    C. Both A and B
    D. Neither A nor B

**14**

18. When floor beams have burned at their ends where they enter a party wall, fire personnel should _____.
    A. Check the far side of the wall to see whether the fire or water has come through
    B. Flush the voids in the wall with water
    C. Check for fire within the insulation materials within the wall
    D. All of the above

19. Which of the following statements about preserving evidence is correct?
    A. Burned paper and ash may be protected by carefully placing them inside glass bottles containing cotton.
    B. Wood suspected of containing paraffin or oil should be wrapped in fresh newspaper and clearly labeled.
    C. Volatile liquids, oil samples, and oil-soaked rags should be put in tin cans and sealed.
    D. None of the above are correct.

20. After fire department personnel have taken the necessary measures to preserve evidence, they should _____.
    A. Remove charred materials
    B. Notify occupant that it is now okay for the occupant or building representative to begin cleaning and pickup operations
    C. Shovel debris into containers and dump debris onto the street rather than onto the yard
    D. All of the above

## IDENTIFICATION

**Identify the following items.**

21. Sprinkler kit

_____

_____

22. Carryall

_____

_____

23. Catchall

_____

_____

**14**

**Identify the maintenance procedures for the following salvage equipment and tools.**

24. Canvas salvage covers _____

_____

_____

_____

_____

25. Synthetic salvage covers _____

_____

_____

_____

_____

26. Water vacuum _____

_____

27. Mops _____

_____

28. Buckets and tubs _____

_____

29. Tools _____

_____

30. Brooms _____

_____

**Identify examples of how firefighters can detect hidden fires in each of the following ways.**

31. By sound _____

_____

32. By touch _____

_____

33. By sight _____

_____

34. By using electronic sensors _____

_____

**14**

**Provide the requested information.**

35. Identify the two things that firefighters must keep in mind regarding the protection and preservation of material evidence.

A. _____

_____

B. _____

_____

## LISTING

36. List the three tools suggested to make up a sprinkler kit.

A. _____

B. _____

C. _____

37. List dangerous building conditions, other than those caused by water, of which firefighters should be aware.

_____

_____

_____

_____

_____

_____

_____

_____

_____

_____

_____

_____

## SHORT ANSWER

**Briefly answer each question in your own words.**

38. How should firefighters arrange building contents before covering with salvage covers?

_____

_____

_____

_____

14

39. What are some of the obstacles common for firefighters attempting to perform salvage operations in commercial occupancies?

_____

_____

_____

_____

40. It is important that fire department personnel determine the condition of the building in the area to be searched before starting a search for hidden fires. What are the two important factors that affect the condition of the building?

_____

_____

_____

_____

41. Generally, in what area of the structure do fire department personnel begin overhaul, and when should they begin?

_____

_____

42. When opening a ceiling from below, where should the firefighter be positioned?

_____

_____

43. How can the firefighter expose hidden fires within the frames or casings of doors and windows?

_____

_____

44. To avoid further water damage to a building, how should firefighters extinguish a small burning object uncovered during overhaul? How should they extinguish a larger item such as a piece of furniture or a mattress?

_____

_____

_____

_____

_____

*Courtesy of Tom McCarthy, Chicago, IL.*

# Building
# Construction

# Building Construction |15

**Define each of the following terms.**

1. Load-bearing wall

   _____

   _____

2. Nonload-bearing wall

   _____

   _____

3. Party wall

   _____

   _____

4. Partition wall

   _____

   _____

5. Fire wall

   _____

   _____

## TRUE/FALSE

**Mark each of the following statements true (T) or false (F). Correct each false statement.**

6. ☐ T ☐ F  It is the obligation of only the safety officer to constantly monitor a structure for unsafe conditions.

   _____

   _____

7. ☐ T ☐ F  Wood shake shingles treated with fire retardant are effective in stopping the spread of fire.

   _____

   _____

**15**

8. ☐ T ☐ F   Firefighters can expect steel structural members to fail at temperatures near or above 540°F (280°C).

_____

_____

9. ☐ T ☐ F   Proper vertical ventilation is essential for slowing the spread of fire in buildings with large, open spaces such as warehouses, churches, and theaters.

_____

_____

10. ☐ T ☐ F   The water applied during fire suppression activities can contribute to a structure's weakening.

_____

_____

11. ☐ T ☐ F   One thing common to all types of trusses is that if one member fails, the entire truss is likely to fail.

_____

_____

## MULTIPLE CHOICE

**Circle the letter before the most appropriate response.**

12. Firefighters A and B are discussing fire and collapse in older buildings. Firefighter A says that older buildings are often more likely to collapse than are newer buildings. Firefighter B says that in older structures the wooden structural components may have dehydrated to the point that their ignition temperature has increased. Who is correct?
    A. Firefighter A
    B. Firefighter B
    C. Both A and B
    D. Neither A nor B

13. Which of the following statements about building collapse is _not_ correct?
    A. Damage caused by the fire may cause collapse.
    B. Damage to the structural system of the building resulting from fire fighting operations may cause collapse.
    C. The longer a fire burns in a building, the more likely it is that the building will collapse.
    D. Buildings of heavy timber construction will likely collapse more quickly than will those of metal truss construction.

**15**

14. Which of the following statements about buildings of lightweight and truss construction is correct?
    A. Lightweight metal and wood trusses are fire-resistant and are not likely to fail until after about an hour of fire exposure.
    B. Truss-containing buildings exposed to fire conditions for 5 to 10 minutes should not be entered.
    C. Metal trusses will not fail if one member fails, but wooden trusses will.
    D. All of the above are correct.

## IDENTIFICATION

**Identify the following items.**

15. NFPA 220

_____

_____

16. Reinforced concrete

_____

_____

17. Heavy content fire loading

_____

_____

**Identify the basic structural characteristics of the following types of building construction and the primary fire hazards associated with each.**

18. Type I construction _____

_____

_____

_____

_____

_____

_____

_____

_____

_____

_____

_____

_____

**15**

19. Type II construction _____

_____

_____

_____

_____

_____

_____

_____

_____

_____

_____

20. Type III construction _____

_____

_____

_____

_____

_____

_____

_____

_____

_____

_____

_____

21. Type IV construction _____

_____

_____

_____

_____

_____

_____

_____

_____

_____

_____

**15**

22. Type V construction _____
_____
_____
_____
_____
_____
_____
_____
_____
_____
_____

**Identify the effects of fire and fire extinguishment on the following building materials.**

23. Wood _____
_____
_____
_____
_____

24. Masonry _____
_____
_____
_____
_____

25. Cast iron _____
_____
_____
_____
_____

26. Steel _____
_____
_____
_____
_____

**15**

27. Reinforced concrete _____
_____
_____
_____
_____

28. Gypsum _____
_____
_____
_____
_____

29. Glass _____
_____
_____
_____
_____

## LISTING

30. List the conditions of a structure that can contribute to the spread and intensity of a fire.
_____
_____
_____
_____
_____
_____

31. List at least five indicators of potential building collapse.
_____
_____
_____
_____
_____
_____
_____
_____
_____

**SHORT ANSWER**

**Briefly answer each question in your own words.**

32. Why is heavy content fire loading such a critical hazard in commercial and storage facilities?

_____

_____

_____

_____

_____

_____

33. What precautions should firefighters take if they believe that building collapse is imminent or possible?

_____

_____

_____

_____

_____

_____

*Courtesy of Tom McCarthy, Chicago, IL.*

# Firefighters'
# Responsibility In Fire
# Cause Determination

16

# Firefighters' Responsibility In Fire Cause Determination | 16

**Define each of the following terms.**

1. Incendiary device

   _____

   _____

2. Trailer

   _____

   _____

**TRUE/FALSE**

**Mark each of the following statements true (T) or false (F). Correct each false statement.**

3. ☐ T ☐ F  In most jurisdictions, the fire inspector has the ultimate legal responsibility for fire cause determination.

   _____

   _____

4. ☐ T ☐ F  If a fire is of suspicious origin, the incident commander should write a chronological account of important circumstances observed by each of the firefighters.

   _____

   _____

5. ☐ T ☐ F  Once the fire investigator has gathered all the required evidence and information at the scene, thorough overhaul procedures can be performed.

   _____

   _____

6. ☐ T ☐ F  The fire department has the authority to bar access to any building only during fire fighting activities.

   _____

   _____

**16**

7.  ☐ T  ☐ F  When there is evidence of possible arson, at least one person from the fire department should remain at the scene until the investigator arrives.

_____

_____

## MULTIPLE CHOICE

**Circle the letter before the most appropriate response.**

8.  No attempt to cross-examine a potential arson suspect should be made by anyone _except_ _____.
    A.  A trained investigator
    B.  The department fire chief
    C.  A firefighter who first identified the suspect
    D.  Either A or B

9.  Firefighters A and B are discussing investigations at the fire scene. Firefighter A says that a firefighter's statements regarding probable cause of the fire should be made only to the investigator. Firefighter B says that such statements should be made public only with the permission of the investigator and the ranking fire officer or incident commander. Who is correct?
    A.  Firefighter A
    B.  Firefighter B
    C.  Both A and B
    D.  Neither A nor B

10. If an investigator is not immediately available, the premises should be guarded and kept under the control of the _____ until all fire cause evidence has been collected.
    A.  Building owner or occupant
    B.  Police department
    C.  Fire department
    D.  None of the above

## IDENTIFICATION

**Identify the information that can be indicated by the following facts and factors.**

11. Time of day _____

_____

12. Weather _____

_____

13. Man-made barriers _____

_____

14. Color of smoke _____

_____

15. Color of flame _____

_____

16. Unusual odors _____

_____

**Provide the requested information.**

17. In order to properly analyze fire cause, it is necessary to do what three things?

_____

_____

_____

18. Identify indications of the presence of accelerants at the fire scene.

_____

_____

_____

_____

_____

_____

19. Identify ways in which the premises can be secured and protected after firefighters have completed fire fighting operations.

_____

_____

_____

_____

_____

_____

**16**

## LISTING

20. Fire cause is a combination of four factors. List these factors.

A. _____

B. _____

C. _____

D. _____

21. List information items of which firefighters should make note while en route to the fire scene.

_____

_____

_____

_____

## SHORT ANSWER

**Briefly answer each question in your own words.**

22. Firefighters working at the scene have the responsibility of noting everything that could point to the cause of the fire. Why is this important?

_____

_____

_____

_____

23. What could discarded burglary tools and evidence of forcible entry prior to the fire indicate to fire department personnel?

_____

_____

24. When does the fire department's authority to control access to the fire scene end?

_____

_____

25. After fire department personnel have left the scene of a fire, what action must they take in order to reenter the scene?

_____

_____

Courtesy of Tom McCarthy, Chicago IL.

# Fire Alarms And Communications

# Fire Alarms And Communications | 17

## TRUE/FALSE

**Mark each of the following statements true (T) or false (F). Correct each false statement.**

1. ☐ T  ☐ F   Fire alarms may be received from public alerting systems and from private alarm systems.

   _____

   _____

2. ☐ T  ☐ F   The dispatcher monitoring the radio should take the same kind of information that would be taken from a telephone caller.

   _____

   _____

3. ☐ T  ☐ F   Fixed-temperature detectors tend to react more quickly than rate-of-rise detectors, but they are not quite as reliable.

   _____

   _____

4. ☐ T  ☐ F   A central station signaling system is a type of private fire alarm signaling system that signals the alarm to a central station where it is then retransmitted to the fire department.

   _____

   _____

5. ☐ T  ☐ F   Normally, only the incident commander at the fire may order multiple alarms or additional response.

   _____

   _____

# 17 | MULTIPLE CHOICE

**Circle the letter before the most appropriate response.**

6. For municipalities receiving more than 2,500 alarms per year, NFPA standards require at least _____ fully trained communication center operator(s) on duty at all times.
   A. One
   B. Two
   C. Three
   D. Four

7. Firefighters A and B are discussing communications equipment. Firefighter A says that all such equipment should be connected to an auxiliary power supply. Firefighter B says that if public service power is interrupted, fire department communications equipment cannot be used. Who is correct?
   A. Firefighter A
   B. Firefighter B
   C. Both A and B
   D. Neither A nor B

8. The universal frequency for reporting emergencies and the one most commonly monitored by emergency providers is _____.
   A. Citizens band channel 5
   B. Citizens band channel 9
   C. Citizens band channel 12
   D. Citizens band channel 13

9. With a _____, the only information that can be transmitted is the location.
   A. Telephone fire alarm box
   B. Radio fire alarm box
   C. Wired telegraph circuit box
   D. All of the above

10. Firefighters A and B are discussing public alerting systems. Firefighter A says that a radio alarm box may alert a fire department by an audible signal, a visual light indicator, and a printed record. Firefighter B says that some radio alarm boxes have two-way communication capabilities. Who is correct?
    A. Firefighter A
    B. Firefighter B
    C. Both A and B
    D. Neither A nor B

11. Which of the following is *not* a possible function of a private fire alarm signaling system.
    A. To summon the fire department or other organized assistance
    B. To supervise extinguishing systems to ensure their operability when needed
    C. To notify occupants to evacuate the premises
    D. To cause extinguishing systems to cease operation upon fire department notification

**17**

12. Rate-of-rise detectors and fixed-temperature alarm initiating devices are types of _____ devices.
    A. Manually activated
    B. Invisible products-of-combustion
    C. Thermal sensitive
    D. Visible products-of-combustion

13. With which of the following types of signaling systems does the alarm go directly to a panel at a public fire alarm office or fire station?
    A. Proprietary protective signaling system
    B. Remote station protective signaling system
    C. Auxiliary protective signaling system
    D. None of the above

14. Those personnel authorized to transmit on the radio should be specified by _____.
    A. NFPA 1221
    B. The FCC
    C. Local department rules
    D. The dispatcher

## IDENTIFICATION

**Identify the following items.**
15. Communications center

_____

_____

16. Public alerting systems

_____

_____

17. NFPA 1221

_____

_____

18. Enhanced 9-1-1 (E-9-1-1) system

_____

_____

19. Private fire alarm signaling systems

_____

_____

**17**

20. FCC

_____

_____

**Identify the procedures that the public should take in order to report emergencies using the following modes of notification.**

21. Telephone

_____

_____

_____

_____

_____

_____

_____

_____

_____

22. Fire alarm pull box

_____

_____

_____

_____

23. Local alarm box

_____

_____

_____

_____

**Provide the requested information.**

24. Identify the two most common methods of sounding an evacuation signal.

A. _____

_____

B. _____

_____

**LISTING**

25. List procedures to follow when processing business calls.

_____
_____
_____
_____
_____
_____
_____
_____
_____
_____
_____
_____
_____
_____
_____
_____
_____

26. List the six basic types of alarm initiating devices.

A. _____
B. _____
C. _____
D. _____
E. _____
F. _____

27. List at least five guidelines regarding radio transmission that firefighters observe.

_____
_____
_____
_____
_____
_____
_____
_____
_____
_____

**17**

28. List information that the firefighter should include when giving a report of conditions upon arrival.

_____

_____

_____

_____

_____

_____

29. List information that should be included when reporting progress to the communications center.

_____

_____

_____

_____

_____

_____

_____

## SHORT ANSWER

**Briefly answer each question in your own words.**

30. When it is necessary to broadcast emergency traffic over the radio, what should the person transmitting do, and what should the dispatcher do as a result of this?

_____

_____

_____

_____

_____

31. When are evacuation signals used?

_____

_____

_____

_____

# 18

# Fire Prevention
And Public
Fire Education

# Fire Prevention And Public Fire Education

## DEFINITIONS OF KEY TERMS

**Define each of the following terms.**

1. Plot plan

_____

_____

2. Fire hazard

_____

_____

3. Personal hazards

_____

_____

## TRUE/FALSE

**Mark each of the following statements true (T) or false (F). Correct each false statement.**

4. ☐ T  ☐ F  The company officer should contact the building occupant prior to an inspection to set a suitable day and time for the inspection.

_____

_____

5. ☐ T  ☐ F  When making an inspection, firefighters should enter the premises at the main entrance but do not need to obtain permission to make the inspection.

_____

_____

6. ☐ T  ☐ F  Firefighters must record every fire inspection, but it is not always necessary for them to make a formal report.

_____

_____

**18**

7. ☐ T   ☐ F   Firefighters must test fire protection systems in the occupancies they inspect.

_____

_____

8. ☐ T   ☐ F   Photoelectric smoke detectors respond slightly faster to smoldering fires than do ionization detectors.

_____

_____

9. ☐ T   ☐ F   Smoke detector batteries should be changed at least once a year.

_____

_____

10. ☐ T   ☐ F   A test button on a smoke detector can only test the device's horn circuit.

_____

_____

11. ☐ T   ☐ F   During school fire exit drills, emphasis should be placed upon orderly evacuation under proper discipline rather than upon speed.

_____

_____

12. ☐ T   ☐ F   All fire exit drill alarms should be sounded on the signal system used to dismiss classes.

_____

_____

## MULTIPLE CHOICE

**Circle the letter before the most appropriate response.**

13. In most municipalities, who is ultimately responsible for fire prevention?
    A. The safety officer
    B. The fire department chief
    C. The police chief
    D. None of the above

18

14. Firefighters A and B are discussing pre-incident planning. Firefighter A says that the success of such planning depends on the firefighters' ability to perform adequate building inspections. Firefighter B says that it depends on their ability to write complete and accurate reports of their findings. Who is correct?
    A. Firefighter A
    B. Firefighter B
    C. Both A and B
    D. Neither A nor B

15. In order to perform routine building inspections, the firefighter must first meet the objectives found in _____.
    A. NFPA 1031
    B. NFPA 1035
    C. NFPA 1500
    D. None of the above

16. Firefighters A and B are discussing building inspections. Firefighter A says that firefighters do not have to obtain permission to conduct an inspection although obtaining such helps create goodwill within the community. Firefighter B says that a representative of the occupant should accompany firefighters during the entire inspection. Who is correct?
    A. Firefighter A
    B. Firefighter B
    C. Both A and B
    D. Neither A nor B

17. If a floor plan was made during a previous building inspection, the firefighter making the new inspection should _____.
    A. Disregard the old floor plan
    B. Obtain a new floor plan from the building occupant/owner
    C. Record any changes that have been made, and change the floor plan accordingly
    D. All of the above

18. Which of the following is an example of a fuel supply hazard?
    A. Dusts such as grain, wood, metal, or coal dust
    B. Metals such as magnesium, sodium, or potassium
    C. Flammable and combustible liquids such as gasoline, oils, lacquers, or alcohol
    D. All of the above

19. Which of the following is *not* an objective of home fire inspection programs?
    A. To obtain proper life safety conditions
    B. To ensure installation of fire protection systems
    C. To keep fires from starting
    D. To help the owner or occupant understand and improve existing conditions

**18**

20. Firefighters A and B are discussing chimneys. Firefighter A says that chimneys supported on wooden posts are likely to crack and allow sparks to set fire to woodwork. Firefighter B says that chimneys should be equipped with spark arrestors to prevent sparks from exiting the chimney. Who is correct?
    A. Firefighter A
    B. Firefighter B
    C. Both A and B
    D. Neither A nor B

21. Tall grass and vegetation should be kept at least _____ from the house.
    A. 10 feet (3 m)
    B. 20 feet (6 m)
    C. 30 feet (10 m)
    D. 40 feet (13 m)

22. Which of the following statements about flammable liquid storage is correct?
    A. Flammable liquids may be kept inside the dwelling only if stored in a safety-type can.
    B. Flammable liquids can be kept in any type of container if stored in an outside storage area.
    C. Flammable liquids should be stored in a safety-type can in an outside storage area.
    D. Both A and C are correct.

23. Which of the following statements about gas appliances is *not* correct?
    A. Every gas appliance is equipped with an automatic gas control device that cuts off the supply of gas when the pilot flame is extinguished.
    B. A separate shutoff should be provided on the supply line to every appliance.
    C. A shutoff should be accessible without having to move the appliance.
    D. Gas leaks can result if corroded piping and rubber tubing are present.

24. Which of the following statements about ionization smoke detectors is *not* correct?
    A. They use small amounts of radioactive material.
    B. When smoke enters the sensing chamber, it causes the electrical current to be reduced, triggering the alarm.
    C. They respond slightly slower to open flaming fires than do photoelectric detectors.
    D. They can respond to particles that are too small to be seen with the human eye.

25. Firefighters A and B are discussing photoelectric smoke detectors. Firefighter A says that a photoelectric detector is activated when light reflecting from smoke particles strikes a photocell. Firefighter B disagrees and says that the detector is activated when smoke obscures the photocell, preventing light from striking it. Who is correct?
    A. Firefighter A
    B. Firefighter B
    C. Both A and B
    D. Neither A nor B

## IDENTIFICATION

**18**

**Identify the following items.**

26. NFPA 1031

_____

_____

27. NFPA 1035

_____

_____

28. NFPA 31

_____

_____

29. Common fire hazard

_____

_____

30. Special fire hazard

_____

_____

**Identify examples of special fire hazards associated with each of the following types of occupancies.**

31. Public assembly occupancies

_____

_____

_____

_____

32. Manufacturing occupancies

_____

_____

_____

_____

**18**

33. Commercial occupancies

_____
_____
_____
_____
_____

**Provide the requested information.**

34. What are some examples of common hazards?

_____
_____
_____
_____

35. When performing a dwelling inspection, which rooms or areas should be inspected?

_____
_____

## LISTING

36. List six reasons why firefighters conduct building inspections.

A. _____
_____

B. _____
_____

C. _____
_____

D. _____
_____

E. _____
_____

F. _____
_____

37. List at least six items the firefighter should observe or check before actually entering the occupancy to be inspected.

_____
_____
_____
_____

**18**

_____
_____
_____
_____
_____
_____
_____
_____
_____
_____
_____

38. List the items of information that should be included in a formal inspection report.

_____
_____
_____
_____
_____
_____
_____
_____
_____
_____
_____
_____

39. List the fire protection systems in occupancies that firefighters should inspect.

_____
_____
_____
_____

40. List items which firefighters should check when inspecting standpipes.

_____
_____
_____
_____
_____
_____
_____
_____

**18**

41. List at least four guidelines for making a successful dwelling inspection.

_____

_____

_____

_____

_____

_____

42. List the five most common causes of fires.

A. _____

B. _____

C. _____

D. _____

E. _____

43. List at least five basic home fire safety rules.

_____

_____

_____

_____

_____

_____

_____

_____

_____

## SHORT ANSWER

**Briefly answer each question in your own words.**

44. What is the difference between pre-incident planning and a building inspection?

_____

_____

_____

_____

_____

_____

45. When should fire department personnel make an inspection follow-up visit?

_____

_____

46. While sketching the floor plan of a building being inspected, when is it necessary for the firefighter to provide a cross-sectional or cutaway view of a particular portion of a building?

_____

_____

47. What is the difference between a report and a record?

_____

_____

_____

_____

48. What should the firefighter check when examining building exits during an inspection?

_____

_____

_____

_____

_____

_____

_____

_____

49. Why do the personal visits by firefighters to perform dwelling inspections generally result in improved community support of the fire department?

_____

_____

_____

_____

_____

50. What is the purpose of a dwelling inspection campaign?

_____

_____

**18**

51. How should people be instructed to behave if their clothing catches fire? Why is it important to impress upon them that they should not run?

_____

_____

_____

_____

52. What are some advantages of battery-operated smoke detectors over those that operate on household current?

_____

_____

53. Where should smoke detectors be placed?

_____

_____

_____

_____

54. What is the primary purpose of school fire exit drills?

_____

_____

# 19

# Firefighter Safety

# Firefighter Safety |19

**Define each of the following terms.**

1. Accident

_____

_____

2. Injury

_____

_____

3. Law

_____

_____

4. Standards

_____

_____

5. Goal

_____

_____

6. Objective

_____

_____

7. Inverter

_____

_____

8. Generator

_____

_____

# 19

**TRUE/FALSE**

**Mark each of the following statements true (T) or false (F). Correct each false statement.**

9. ☐ T ☐ F   A fire department is not required to meet the requirements set forth in NFPA 1500 unless that standard has been legally adopted by the authority having jurisdiction.

_____

_____

10. ☐ T ☐ F   A physical fitness program will improve general health but will not help reduce stress-related injuries.

_____

_____

11. ☐ T ☐ F   To improve the efficiency of their cardiovascular systems, firefighters should train one or two times a week at a level sufficient to raise their pulse rates within a specific range.

_____

_____

12. ☐ T ☐ F   It is important that physical fitness programs include both cardiovascular training and training to strengthen muscles and joints.

_____

_____

13. ☐ T ☐ F   A firefighter should not lift a large load that he or she cannot lift by holding the load close to the body and lifting with the leg muscles.

_____

_____

14. ☐ T ☐ F   A person who smokes cigarettes has a significantly greater chance of developing lung cancer than does a nonsmoker, but a smoker is no more likely to have a heart attack than is a nonsmoker.

_____

_____

15. ☐ T ☐ F   The effects of smoking can directly affect a firefighter's ability to physically perform at a safe and effective level.

_____

_____

**19**

16. ☐ T  ☐ F  The greatest number of firefighter injuries and fatalities occur at the emergency scene.

_____

_____

17. ☐ T  ☐ F  Excitement or disorientation causes firefighters to expend their air supply faster than normal.

_____

_____

18. ☐ T  ☐ F  Natural gas service to a building may be halted by turning the petcock on the gas meter so that it is parallel to the pipe.

_____

_____

19. ☐ T  ☐ F  Water service to a building may be stopped by closing the valve on the feeder main to the building.

_____

_____

20. ☐ T  ☐ F  Not only can a "cheater" overload a tool and cause it to break suddenly, but it can also cause the weakened tool to break later when being used normally.

_____

_____

21. ☐ T  ☐ F  With an unenclosed jump seat, safety bars are more effective than safety gates.

_____

_____

22. ☐ T  ☐ F  Firefighters should use handrails when mounting or dismounting the apparatus unless the apparatus has an aerial device extended close to electrical wires.

_____

_____

23. ☐ T  ☐ F  A department's safety officer should examine station conditions and work procedures as closely as fireground procedures.

_____

_____

# 19 | MULTIPLE CHOICE

**Circle the letter before the most appropriate response.**

24. Firefighters A and B are discussing proper nutrition. Firefighter A says that complex carbohydrates should account for about 48 percent of caloric intake. Firefighter B says that overall fat consumption should be about 10 percent of total calories consumed. Who is correct?
    A. Firefighter A
    B. Firefighter B
    C. Both A and B
    D. Neither A nor B

25. Firefighters A and B are discussing cigarette smoking. Firefighter A says that the firefighter who smokes is more prone to lung injury than the average citizen because of double exposure by cigarette smoke and products of combustion found on the fire scene. Firefighter B says that the protective mechanisms of the lungs fail after prolonged exposure to toxic substances contained in tobacco smoke and that they cannot be restored. Who is correct?
    A. Firefighter A
    B. Firefighter B
    C. Both A and B
    D. Neither A nor B

26. Which of the following statements about fire officers is *not* correct?
    A. All fire officers should meet the requirements specified in NFPA 1021.
    B. Officers should be trained to evaluate the safety of structures.
    C. Officers should rely on the incident commanders for incident evaluations.
    D. Officers should be trained to deploy equipment and personnel.

27. Which of the following ICS functional areas has the authority to order and release resources?
    A. Command
    B. Operations
    C. Planning
    D. Logistics

28. Generally, the _____ ICS area will be of concern only at large-scale, long-term incidents.
    A. Operations
    B. Logistics
    C. Planning
    D. Finance

29. Which of the following is the safest way for firefighters to disconnect electrical service to a building?
    A. Cutting the drop wires into the building
    B. Pulling the electrical meters
    C. Shutting off the main breakers at the electrical service box or panel
    D. Disconnecting electrical wires from the nearest utility pole

30. Firefighters A and B are discussing restoring water, gas, and electrical service to a building. Firefighter A says that firefighters can turn on water and gas service but should let the electrical company turn back on the electrical power to the building. Firefighter B says that firefighters should return any of the three utilities to service if previously disconnected. Who is correct?
    A. Firefighter A
    B. Firefighter B
    C. Both A and B
    D. Neither A nor B

31. According to the applicable OSHA standard, aerial devices or ground ladders shall be kept a minimum of _____ from lines rated 50 kV or lower.
    A. 5 feet (1.5 m)
    B. 10 feet (3 m)
    C. 15 feet (4.5 m)
    D. 50 feet (15 m)

32. Firefighters A and B are discussing pole-top fires. Firefighter A says that unless part of the pole is in danger of falling, firefighters should let the fire burn until utility personnel shut down the power. Firefighter B says that a Class C fire extinguisher should be used if necessary. Who is correct?
    A. Firefighter A
    B. Firefighter B
    C. Both A and B
    D. Neither A nor B

33. Which of the following guidelines would most likely *not* be included in a typical departmental policy for handling electrical exposure?
    A. As soon as they arrive at the scene, firefighters should cut wires to the structure to shut off all power.
    B. Firefighters should not use solid or straight hose streams when a possible electrical hazard exists.
    C. Firefighters should avoid parking apparatus under overhead wires when possible.
    D. Firefighters should wear full protective clothing when electrical hazards exist.

34. Design, construction, testing, and use of eye and face protection must be in accordance with _____.
    A. NFPA 1001
    B. ANSI Z98.9
    C. NFPA 1500
    D. ANSI Z87.1

35. Which of the following are guidelines for the servicing and maintenance of power plants and lighting equipment?
    A. Check fluid levels weekly; check gas and oil levels after every use.
    B. Test electrical devices for operating status while the power plant is running.
    C. Run power plants at least once a week, for at least 20 minutes, while powering an electrical device.
    D. All of the above

**19**

36. Which of the following statements is *not* correct?
    A. Firefighters should be seated with their seat belts fastened when the vehicle is in motion unless they are riding in an enclosed cab.
    B. Firefighters that are not riding in enclosed seats should wear helmets and eye protection.
    C. Firefighters riding on an apparatus should wear hearing protection if sirens and noise levels exceed 90 decibels.
    D. Firefighters should not ride on the tailboard of the apparatus.

37. NFPA 1500 requires that all personnel who may engage in structural fire fighting participate in training at least _____.
    A. Twice a month
    B. Monthly
    C. Every two months
    D. Twice a year

38. Firefighters A and B are discussing tools and equipment used for training evolutions. Firefighter A says that items used frequently for training will often wear out sooner than those used routinely in the fire station. Firefighter B says that all tools and equipment should be inspected before each drill. Who is correct?
    A. Firefighter A
    B. Firefighter B
    C. Both A and B
    D. Neither A nor B

## IDENTIFICATION

**Identify the following items.**

39. NFPA 1500

_____

_____

40. NFPA 1021

_____

_____

41. ICS

_____

_____

42. Support branch

_____

_____

43. Service branch

_____

_____

44. GFCI

_____

_____

**Identify the responsibilities associated with each of the five major functional areas of the Incident Command System.**

45. Command _____

_____

_____

46. Operations _____

_____

_____

47. Planning _____

_____

_____

48. Logistics _____

_____

_____

49. Finance _____

_____

_____

**Provide the requested information.**

50. Identify actions that a disoriented firefighter should take.

_____

_____

_____

_____

_____

_____

_____

_____

_____

19

# 19

## LISTING

51. List at least five guidelines to observe when using hand and power tools.

_____
_____
_____
_____
_____
_____
_____
_____
_____

52. List power saw safety rules.

_____
_____
_____
_____
_____
_____
_____
_____
_____
_____
_____
_____
_____

## SHORT ANSWER

**Briefly answer each question in your own words.**

53. When using a personnel accountability system, where should copies of the on-duty list be kept?

_____
_____
_____
_____

54. How is the simple tag system used for personnel accountability?

_____

_____

_____

_____

_____

_____

_____

_____

_____

_____

_____

_____

55. How does an SCBA tag system differ from a simple tag system?

_____

_____

_____

_____

_____

_____

_____

_____

_____

_____

_____

_____

_____

_____

56. During an emergency, power company personnel should handle energized electrical equipment. What is the exception to this policy?

_____

_____

_____

_____

_____

**19**

**19**

57. Why is it necessary for a firefighter to exercise care when using portable floodlights?

_____

_____

_____

_____

58. Why should firefighters match the amount of lighting with the amount of power available from the power plant?

_____

_____

_____

_____

59. At the fire station, what are the two most common types of accidents that result in injury?

_____

_____

# Answers

# Chapter 1 Answers

## DEFINITIONS OF KEY TERMS

1. *Combustion* — The self-sustaining process of rapid oxidation of a fuel, which produces heat and light (5)
2. *Fire* — The result of a rapid combustion reaction (5)
3. *British thermal unit (Btu)* — The amount of heat needed to raise the temperature of one pound of water one degree Fahrenheit (5)
4. *Calorie* — The amount of heat needed to raise the temperature of one gram of water one degree Celsius (6)
5. *Celsius (Centigrade)* — Metric unit of temperature measurement; 0 is the melting point of ice, and 100 degrees is the boiling point of water (6)
6. *Fahrenheit* — Unit of temperature measurement primarily used in the United States; 32 degrees is the melting point of ice, and 212 degrees is the boiling point of water (6)
7. *Fire point* — The temperature at which a liquid fuel will produce vapors sufficient to support continuous combustion once ignited; the fire point is usually a few degrees above the flash point (6)
8. *Flame spread* — The movement of flame away from the ignition source (6)
9. *Flash point* — The minimum temperature at which a liquid fuel gives off sufficient vapors to form an ignitable mixture with the air near the surface; at this temperature, the ignited vapors will flash but will not continue to burn (6)
10. *Heat* — The form of energy that raises temperature (6)
11. *Ignition temperature* — The minimum temperature to which a fuel in air must be heated to start self-sustained combustion without a separate ignition source (6)
12. *Oxidation* — The complex chemical reaction of organic materials with oxygen or other oxidizing agents resulting in the formation of more stable compounds (6)
13. *Conduction* — The transfer of heat from one body to another by direct contact of the two bodies or by an intervening heat-conducting medium (9)
14. *Convection* — The transfer of heat by the movement of air or liquid (9)
15. *Radiation* — The transfer of heat by heat waves (10)
16. *Pyrolysis* — The chemical decomposition of a substance through the action of heat (10)
17. *Specific gravity* — The density of liquids in relation to water (12)
18. *Vapor density* — The density of gas or vapor in relation to air (12)

## TRUE/FALSE

19. True (8)
20. True (8)
21. False. Of the three states of matter in which fuel may be found, *only gases* burn. (10)
22. True (10)
23. False. If a solid fuel is in a *vertical* position, fire spread will be more rapid than if it is in a *horizontal* position. (11)
24. False. If heat is dissipated faster than it is generated, a *negative* heat balance is created; a positive heat balance is required to maintain combustion. (13)

**1**

25. False. *Heat* is the product of combustion that is responsible for the spread of fire. (19)
26. True (20)
27. True (20)

## MATCHING

28. B (7)
29. E (8)
30. D (7)
31. A (7)
32. C (7)

## IDENTIFICATION

33. *Boiling point* — The temperature of a substance when the vapor pressure exceeds atmospheric pressure. At this temperature, the rate of evaporation exceeds the rate of condensation. At this point, more liquid is turning into gas than gas is turning back into a liquid. (5)
34. *Flammable or explosive limits* — The percentage of a substance (vapor) in air that will burn once it is ignited. Most substances have an upper (too rich) and a lower (too lean) flammable limit. (6)
35. *Vapor pressure* — A measure of the tendency of a substance to evaporate (6)
36. *The Law of Heat Flow* — A natural law of physics that specifies that heat tends to flow from a hot substance to a cold substance (9)
37. *Radiative feedback* — Radiant heat providing energy for continued vaporization (13)
38. A. *Heat of combustion* — The amount of heat generated by the combustion (oxidation) reaction (6)
    B. *Spontaneous heating* — The heating of an organic substance without the addition of external heat (7)
    C. *Heat of decomposition* — The release of heat from decomposing compounds, usually due to bacterial action (7)
    D. *Heat of solution* — The heat released by the solution of matter in a liquid (7)
39. A. *Incipient phase* — The earliest phase of a fire beginning with the actual ignition. The fire is limited to the original materials of ignition; the oxygen content in the air has not been significantly reduced; and the fire is producing water vapor ($H_2O$), carbon dioxide ($CO_2$), and perhaps a small quantity of sulfur dioxide ($SO_2$), carbon monoxide (CO), and other gases. Flame temperature may be well above 1,000°F (537°C), yet temperature in the room may be only slightly increased. (15)
    B. *Steady-state burning phase* — Phase of fire when sufficient oxygen and fuel are available for fire growth and open burning to a point where total involvement is possible. Heated gases spread out laterally from the top downward, eventually igniting all the combustible material in the upper levels of the room. Temperature in the upper regions can exceed 1,300°F (700°C). "Clear burning" can be possible, and flashover can occur during this phase. (16)

C. *Hot-smoldering phase* — Phase of fire when flames may cease to exist if the area of confinement is sufficiently airtight. Burning is reduced to glowing embers, air pressure from gases being given off may build to the extent that smoke and gases are forced through small cracks, and room temperatures in excess of 1,000°F (537°C) are possible. A backdraft is possible in this phase. (17)

1

## LISTING

40. Answer should include at least five of the following:
    - Pressurized smoke exiting small openings
    - Black smoke becoming dense gray yellow
    - Confinement and excessive heat
    - Little or no visible flame
    - Smoke leaving the building in puffs or at intervals
    - Smoke-stained windows
    - Muffled sounds
    - Sudden rapid movement of air inward when opening is made (17-18)
41. A. Heat
    B. Flame
    C. Smoke
    D. Fire gases (19)
42. Ordinary combustible materials such as wood, cloth, paper, rubber, and many plastics (21)
43. Flammable and combustible liquids and gases such as gasoline, oil, lacquers, paints, mineral spirits, and alcohols (21)
44. Energized electrical equipment such as household appliances, computers, transformers, and overhead transmission lines (22)
45. Combustible metals such as aluminum, magnesium, titanium, zirconium, sodium, and potassium (22)

## SHORT ANSWER

46. That because their vapor density is greater than one, these hydrocarbons are heavier than air and will sink and hug the ground and will flow into low lying areas (12)
47. Fuel, temperature, oxygen, and the uninhibited chemical chain reaction necessary for the flaming mode of combustion (13)
48. The fire triangle represents the surface or smoldering mode of combustion; its three sides represent fuel, temperature, and oxygen. The fire tetrahedron represents the flaming mode of combustion; three of its four sides represent the same three items as the fire triangle, and the fourth side represents the uninhibited chemical chain reaction. (13)
49. 21 percent. Oxygen concentrations below 21 percent can threaten life safety. From a fire standpoint, the intensity of a fire will begin to decrease when the oxygen level falls below 18 percent; oxygen concentrations below 15 percent will not support combustion. (13-14)

**1**

50. Rollover occurs when superheated gases accumulated at the ceiling level mix with oxygen and, when they reach their flammable range, ignite, developing a fire front that rapidly expands and rolls over the ceiling. Flashover occurs when all of the contents of the fire area are heated to their ignition temperatures and then simultaneously ignite; flames flash over the entire surface of the room or area. Rollover differs from flashover in that only the gases are burning and not the contents of the room. (15, 16)

51. The tendency of gases to form into layers, according to temperature. It is important not to disrupt this layering because as long as the hottest air and gases are allowed to rise, the lower levels will be safer for firefighters. In addition, the normal layering of the hottest gases to the top and out the ventilation opening can be disrupted if water is improperly applied. (18)

52. Water is used in a cooling or quenching effect to reduce the temperature of the burning material below its ignition temperature. Class A foams may be used to extinguish Class A fires, particularly those that are deep seated in bulk materials. (21)

53. The smothering or blanketing effect of oxygen exclusion is most effective for extinguishment. Other extinguishing methods include removal of fuel and temperature reduction when possible. (21-22)

54. A nonconducting extinguishing agent such as Halon, dry chemical, or carbon dioxide is used for extinguishment. The safest extinguishment procedure is to first deenergize high voltage circuits and then to treat the fire as a Class A or Class B fire depending upon the fuel involved. (22)

55. Special extinguishing agents available for control of fire for the different combustible metals are used for extinguishment by covering up the burning material and smothering the fire. (22)

**2**

# Chapter 2 Answers

**TRUE/FALSE**

1. False. Portable fire extinguishers for *Class A* and *Class B* fires have a letter rating and a numerical rating; those for *Class C* and *Class D* fires have only a letter rating. (25)

2. False. The effectiveness of an extinguisher rated for a Class D fire is indicated on *the faceplate* of the extinguisher. (25)

3. True (29)

4. True (29)

5. False. Modern fire extinguishers are designed to be carried to the fire in an upright position and then *operated in an upright position*. (32)

6. True (38)

7. False. Leaking, corroded, or otherwise damaged extinguisher shells or cylinders should be *discarded or returned to the manufacturer for repair*. (42)

8. False. Servicing portable fire extinguishers is the responsibility of the *property owner or building occupant*. (43)

9. True (44)

## MULTIPLE CHOICE

10. B (25)
11. C (25)
12. A (25)
13. C (27)
14. B (42)

## MATCHING

15. A (26, 29)
16. D (26, 29)
17. C (26, 29)
18. B (26, 29)
19. B (26)
20. A (26)
21. D (26)
22. C (26)

## IDENTIFICATION

23. *NFPA 10 — Standard for Portable Fire Extinguishers* (25)
24. *Underwriter's Laboratories, Inc. (UL) and Underwriter's Laboratories of Canada (ULC) —* Entities that conduct tests on portable fire extinguishers to determine their numerical ratings (25)
25. *AFFF —* Aqueous film forming foam (34)
26. *Halon 1211 —* Bromochlorodifluoromethane (35)
27. *Halon 1301 —* Bromotrifluoromethane (35)
28. *Dry chemical agents —* Extinguishing agents used on Class A-B-C fires and/or Class B-C fires (38, 39)
29. *Dry powder agents —* Extinguishing agents used on Class D fires only (39)
30. *P —* Pull the pin at the top of the extinguisher that keeps the handle from being pressed.
    *A —* Aim the nozzle or outlet toward the fire.
    *S —* Squeeze the handle above the carrying handle to discharge the agent.
    *S —* Sweep the nozzle back and forth at the base of the flames to disperse the extinguishing agent. Make sure that the fire is out. (30-31)
31. *Pump tank water extinguishers*
    A. *Class of fire suited for —* Class A
    B. *Size —* 1½ to 5 gallons (6 L to 20 L)
    C. *Stream reach —* 30 to 40 feet (9 m to 12 m)
    D. *Discharge time —* 45 seconds to 3 minutes
    E. *Basic operation —* These extinguishers are equipped with pumps that the user must pump to deliver water. (32)

**2**

32. *Stored-pressure water extinguishers*
    A. *Class of fire suited for* — Class A
    B. *Size* — 1¼ to 2½ gallons (5 L to 10 L)
    C. *Stream reach* — 30 to 40 feet (9 m to 12 m)
    D. *Discharge time* — 30 to 60 seconds
    E. *Basic operation* — These extinguishers use compressed air to discharge the water. The user must release the shutoff device (squeeze the handle) to expel water. (33-34)

33. *Aqueous film forming foam (AFFF) extinguishers*
    A. *Class of fire suited for* — Class A and Class B
    B. *Size* — most common size is 2½ gallons (10 L)
    C. *Stream reach* — 20 to 25 feet (6 m to 7.5 m)
    D. *Discharge time* — 50 seconds
    E. *Basic operation* — These extinguishers use compressed air to discharge the water/AFFF concentrate mixture. The user must release the shutoff device to expel the mixture and should apply the foam using a side-to-side sweeping motion across the entire width of the fire. (34-35)

34. *Halon 1211 extinguishers*
    A. *Class of fire suited for* — Class B and Class C (Those extinguishers having greater than a 9-pound [4 kg] capacity will also have a small Class A rating.)
    B. *Size* — 2½ to 22 pounds (1 kg to 10 kg); larger wheeled units, up to 150 pounds (68 kg)
    C. *Stream reach* — 8 to 18 feet (2.5 m to 5.5 m)
    D. *Discharge time* — 8 to 18 seconds
    E. *Basic operation* — These extinguishers contain Halon 1211 stored as a liquefied compressed gas, and nitrogen gives the Halon 1211 added pressure when discharged. The user must squeeze the shutoff handle to expel the agent, which is released in a clear liquid stream, and should direct the discharge at the base of the flame, first at the near edge and gradually progressing forward while moving from side to side. (35)

35. *Halon 1301 extinguishers*
    A. *Class of fire suited for* — Class B and Class C
    B. *Size* — 2½ pounds (1 kg)
    C. *Stream reach* — 4 to 6 feet (1.3 m to 2 m)
    D. *Basic operation* — These extinguishers discharge Halon 1301 in a nearly invisible gaseous form. The user must squeeze the shutoff handle to expel the agent. (35-36)

36. *Carbon dioxide extinguishers (hand carried)*
    A. *Class of fire suited for* — Class B and Class C
    B. *Size* — 2 to 20 pounds (1 kg to 9 kg)
    C. *Stream reach* — 3 to 8 feet (1 m to 2.5 m)
    D. *Discharge time* — 8 to 30 seconds
    E. *Basic operation* — These extinguishers store carbon dioxide as a liquefied compressed gas and discharge it in a gaseous form to smother the fire. The user must squeeze the shutoff handle to expel the agent and should point the discharge horn at the base of the fire. On flammable liquid fires, the user should discharge the agent first at the near edge of the fire, gradually progressing forward, applying the agent with a slow side-to-side motion. (36-37)

37. *Carbon dioxide wheeled units*
    A. *Class of fire suited for* — Class B and Class C
    B. *Size* — 50 to 100 pounds (23 kg to 45 kg)
    C. *Stream reach* — 8 to 20 feet (2 m to 6 m)
    D. *Discharge time* — 30 seconds
    E. *Basic operation* — These extinguishers store carbon dioxide as a liquefied compressed gas and discharge it in a gaseous form to smother the fire. The user must unreel the extinguisher hose before releasing the agent. (37-38)
38. *Dry chemical extinguishers (hand carried)*
    A. *Class of fire suited for* — ordinary: Class B and Class C; multipurpose: Class A, Class B, and Class C
    B. *Size* — 2½ to 30 pounds (1.5 kg to 11 kg)
    C. *Stream reach* — 5 to 20 feet (2 m to 6 m)
    D. *Discharge time* — 10 to 25 seconds
    E. *Basic operation* — The stored-pressure type extinguisher stores the agent under pressure; the user must pull the pin and squeeze the shutoff handle to discharge the agent. The cartridge-operated extinguisher has a separate agent tank and pressure cylinder; the user must first remove the hose from its storage position, point the top of the extinguisher away from people, and then depress the activation plunger to charge the tank. The user can then discharge the agent by squeezing the control handle on the nozzle. With either type, the user should apply the agent in a side-to-side sweeping motion, attacking the near edge of the fire and progressing forward. (38-40)
39. *Dry chemical wheeled units*
    A. *Class of fire suited for* — Class A, Class B, and Class C
    B. *Size* — 75 to 350 pounds (34 kg to 160 kg)
    C. *Stream reach* — up to 45 feet (14 m)
    D. *Discharge time* — 20 to 120 seconds
    E. *Basic operation* — With this extinguisher, the extinguishing agent is kept in one tank and the pressurizing gas is stored in a separate cylinder. The user must remove the hose from the storage position, turn a handwheel or activate a quick-pressurization device to release the gas into the agent tank, allow the tank to pressurize, and then discharge the agent by squeezing the handle. On Class A fires, the user should direct the agent at the burning surfaces to cover them with chemical. (40)

## LISTING

40. A. Wood crib
    B. Wood panel
    C. Excelsior (25)
41. The following are correct:
    - Reactions between the metal and the agent
    - Toxicity of the agent
    - Toxicity of the fumes produced and the products of combustion
    - The possible burnout of the metal instead of extinguishment (29)

**2**

42. The following are correct:
    - Hazards to be protected
    - Severity of the fire
    - Atmospheric conditions
    - Personnel available
    - Ease of handling extinguisher
    - Any life hazard or operational concerns (30)
43. A. That the extinguisher is in its designated location
    B. That it has not been actuated or tampered with
    C. That there is no obvious physical damage or condition present that will prevent its operation (43)
44. A. Mechanical parts
    B. Extinguishing agents
    C. Expelling means (44)

## SHORT ANSWER

45. According to their intended use on the four classes of fire (A, B, C, and D) (25)
46. The approximate square foot (square meter) area of a flammable liquid fire that a nonexpert operator can extinguish (27)
47. One method uses colored geometric shapes with the class letter shown within the shape. The other method uses a picture-symbol labeling system that indicates with which type of fire the extinguisher is to be used and also emphasizes those on which it is not to be used. (29-30)
48. By adding a compatible antifreeze solution to the water (32)
49. It should be of sufficient depth to adequately cover the fire area and provide a smothering blanket. The agent should be applied gently to avoid breaking any crust that may have formed over the burning metal because the fire may flare up and expose more raw material to combustion. (41)
50. The fire should first be covered with powder, and a 1- or 2-inch (25 mm to 50 mm) layer of powder should be spread out nearby. Then the burning metal should be shoveled onto this layer with more powder added as needed. (41)
51. Because many types are no longer recommended for use. These should be removed from service and replaced with new extinguishers that meet NFPA 10. (42)

# Chapter 3 Answers

## DEFINITIONS OF KEY TERMS

1. *Pulmonary edema* — Accumulation of fluids in the lungs (64)
2. *Synergistic effect* — The phenomenon in which the combined effect of two or more substances is more toxic or more irritating than the total effect of each substance inhaled separately (64)
3. *Open-circuit SCBA* — Breathing apparatus that uses compressed air and vents exhaled air to the outside (73)
4. *Closed-circuit SCBA* — Breathing apparatus that uses compressed or liquid oxygen and keeps the user's exhaled air within the system for reuse (72)
5. *Cascade system* — A series of at least three, 300-cubic-foot (8 490 L) cylinders (99)
6. *Skip breathing* — An emergency breathing technique used to extend the use of the remaining air supply by inhaling, holding the breath as long as it would take to exhale, and then inhaling once again before exhaling (103)

## TRUE/FALSE

7. True (51)
8. True (52)
9. False. Work uniforms that meet NFPA 1975 are designed to be fire resistant and are *not* designed to be worn for fire fighting operations *except under* structural fire fighting clothing. (61)
10. False. When oxygen concentrations are below 18 percent, the human body responds by *increasing* the respiratory rate. (63)
11. False. The tissue damage resulting from inhaling hot air is *not* immediately reversible by introducing fresh, cool air. (64)
12. True (65)
13. True (71)
14. True (71)
15. False. *Lightweight aluminum and combination fiberglass-wrapped aluminum* cylinders are the most common cylinder types and are *10 percent stronger than steel.* (74)
16. False. The SCBA regulator pressure gauge should read within *100* psi (*700* kPa) of the SCBA cylinder gauge if increments are in psi (kPa). (75)
17. True (74)
18. False. When preparing to don the SCBA backpack using the crossed-arms coat method, the firefighter should crouch or kneel at the *valve end of the cylinder.* (82)
19. True (84)
20. True (86)
21. False. The side- or rear-mounted SCBA allows several steps of the donning procedure to be eliminated *but does not permit donning en route.* (86)
22. True (89)
23. True (102)

**3**

24. D (52)
25. C (53)
26. A (56, 57)
27. C (58)
28. C (62)
29. A (64)
30. C (69-70)
31. D (73)
32. B (74)
33. C (74)
34. D (74)
35. A (75)
36. B (78)
37. C (81)
38. C (81)
39. A (89-90, 92)
40. C (104-105)

**IDENTIFICATION**

41. *NFPA 1972 — Standard on Helmets for Structural Fire Fighting* (51)
42. *NFPA 1500 — Standard on Fire Department Occupational Safety and Health Program* (52)
43. *NFPA 1971 — Standard on Protective Clothing for Structural Fire Fighting* (55)
44. *NFPA 1973 — Standard on Gloves for Structural Fire Fighting* (58)
45. *NFPA 1974 — Standard for Protective Footwear for Structural Fire Fighting* (59)
46. *NFPA 1975 — Standard on Station/Work Uniforms for Fire Fighters* (60)
47. *CO* — Carbon monoxide (65)
48. *HCl* — Hydrogen chloride (66)
49. *HCN* — Hydrogen cyanide (67)
50. $CO_2$ — Carbon dioxide (67)
51. $NO_2$ — Nitrogen dioxide (68)
52. $COCl_2$ — Phosgene (68)
53. *Helmet* — Protects the head from impact and puncture injuries as well as from scalding water (49)
54. *Protective hood* — Protects portions of the firefighter's face, ears, and neck not covered by the helmet or coat (50)
55. *Protective coat and trousers* — Protect trunk and limbs against cuts, abrasions, and burn injuries (resulting from radiant heat), and provide limited protection from corrosive liquids (50)
56. *Gloves* — Protect the hands from cuts, wounds, and burn injuries (50)
57. *Safety shoes or boots* — Protect the feet from burn injuries and puncture wounds (50)
58. *Eye protection (goggles or faceshields)* — Protects the wearer's eyes from flying solid particles or liquids (50)

59. *Hearing protection* — Limits noise-induced damage to the firefighter's ears when loud noise situations cannot be avoided (50)
60. *Self-contained breathing apparatus (SCBA)* — Protects the face and lungs from toxic smoke and products of combustion (50)
61. *Personal Alert Safety System (PASS)* — Provides life-safety protection by emitting a loud shriek if the firefighter should collapse or remain motionless for approximately 30 seconds (50)
62. *Carbon monoxide*
    A. *Source* — Incomplete combustion
    B. *Characteristics* — Colorless, odorless gas
    C. *Effects on the body* — Can cause headaches, dizziness, nausea, vomiting, cherry-red skin, unconsciousness; combines with hemoglobin and crowds oxygen from the blood, causing eventual hypoxia of the brain and tissues, resulting in death, if the process is not reversed (65)
63. *Hydrogen chloride*
    A. *Source* — By-product of combustion of plastics, such as PVC, containing chlorine
    B. *Characteristics* — Colorless gas with a pungent odor
    C. *Effects on the body* — Irritates eyes and respiratory tract; causes swelling and obstruction of the upper respiratory tract; labored breathing and suffocation can result (66)
64. *Hydrogen cyanide*
    A. *Source* — Combustion of wool, nylon, polyurethane foam, rubber, and paper materials
    B. *Characteristics* — Colorless gas with an almond odor
    C. *Effects on the body* — Interferes with respiration at the cellular and tissue levels; might cause gasping respirations, muscle spasms, increased heart rate, and sudden collapse (67)
65. *Carbon dioxide*
    A. *Source* — Complete combustion of carboniferous materials
    B. *Characteristics* — Colorless, odorless, and nonflammable
    C. *Effects on the body* — At a 5 percent concentration causes a marked increase in respiration, along with headache, dizziness, sweating, and mental excitement; at higher concentrations can cause death within minutes from paralysis of the brain's respiratory center (67)
66. *Nitrogen dioxide*
    A. *Source* — Commonly formed in silos and grain bins; is also liberated when pyroxylin plastics decompose
    B. *Characteristics* — Reddish-brown gas
    C. *Effects on the body* — Irritates nose and throat; mixes with water and reacts in the presence of oxygen to form nitric and nitrous acids, which are neutralized by alkalies in the body tissues and form nitrites and nitrates, which in turn cause arterial dilation, variation in blood pressure, headaches, and dizziness and can lead to collapse and coma (68)
67. *Phosgene*
    A. *Source* — Produced when refrigerants, such as freon, contact flame
    B. *Characteristics* — Colorless, tasteless gas with a disagreeable musty-hay odor
    C. *Effects on the body* — Causes coughing and irritates eyes and forms hydrochloric acid in the lungs when inhaled; is deadly at 25 ppm (68)

**3**

68. *Limited visibility* — The facepiece reduces peripheral vision, and facepiece fogging can reduce overall vision. (72)

69. *Decreased ability to communicate* — The facepiece hinders voice communication. (72)

70. *Increased weight* — Depending on the model, the protective breathing equipment adds 25 to 35 pounds (11 kg to 16 kg) of weight to the firefighter. (72)

71. *Decreased mobility* — The increase in weight and the splinting effect of the harness straps reduce the firefighter's mobility. (72)

72. *Physical condition of user* — If the wearer is in poor physical condition, the air supply will be expended faster. (72)

73. *Degree of physical exertion* — The harder the firefighter exerts himself or herself, the faster the air supply is expended. (72)

74. *Emotional stability* — A person who becomes excited will increase his or her respirations and use air faster. (72)

75. *Condition of apparatus* — Minor leaks and poor adjustment of regulators result in excess air loss. (72)

76. *Cylinder pressure before use* — If the cylinder is not filled to capacity, the amount of working time will be reduced proportionately. (72)

77. *Training and experience* — Properly trained and highly experienced personnel will be able to draw the maximum air supply from a cylinder (72)

78. A. Turnout boots for fire fighting and emergency activities
    B. Safety shoes for station wear and other fire department activities that include inspections, emergency medical responses, and similar activities (59)

## LISTING

79. A. Statement saying that the garment meets the requirements of the standard
    B. The limitations of the garment
    C. Basic recommendations for proper care of the garment (55)

80. A. Outer shell
    B. Moisture barrier
    C. Thermal barrier (55)

81. Answer should include at least five of the following:
    - Remove dirt from the shell.
    - Remove chemicals, oils, and petroleum products from the shell as soon as possible.
    - Repair or replace helmets that do not fit properly.
    - Repair or replace helmets that are damaged.
    - Check for adequate separation between the suspension web and the outer shell.
    - Inspect suspension systems frequently to detect deterioration. Replace if necessary.
    - Consult the helmet manufacturer if a helmet needs repainting.
    - Remove from service and check polycarbonate helmets that have come into contact with hydraulic oil from a rescue tool. (61-62)

82. A. Wash oil, grease, chemicals, and debris from the boot because they deteriorate rubber.
    B. Store rubber boots in a cool, dark place because ozone will deteriorate rubber and cause the boots to lose their protective quality.
    C. Replace worn, cut, or punctured boots that cannot be repaired
    D. Apply the recommended dressing to leather boots at the required intervals. (62)
83. A. Oxygen deficiency
    B. Elevated temperatures
    C. Smoke
    D. Toxic atmospheres (with or without fire) (63)
84. A. Nature of the combustible
    B. Rate of heating
    C. Temperature of the evolved gases
    D. Oxygen concentration (65)
85. A. Backpack and harness assembly
    B. Air cylinder assembly
    C. Regulator assembly
    D. Facepiece assembly (73)
86. A. Releasing cylinder air by quickly opening and closing the bypass valve
    B. Using a nosecup that deflects exhalations away from the lens
    C. Applying an antifogging chemical to the facepiece lens (78)
87. The following are correct:
    - Check the air cylinder gauge to ensure that the cylinder is full.
    - Check the regulator gauge and cylinder gauge to ensure that they read within 100 psi of the same pressure or that the readings are relatively close to each other if gauges are not marked in increments of 100 psi.
    - Check the harness assembly to ensure that it is fully extended.
    - Check the regulator valves to ensure that they are in the proper setting. (81)
88. The following are correct:
    - Check to ensure that the cylinder is at least 90 percent full.
    - Check to ensure that all gauges work. The gauges on the cylinder and regulator should read within 100 psi; if not in increments of 100 psi, the readings should be relatively close to each other.
    - Check to ensure that the low-pressure alarm is in working condition. The alarm should sound briefly when the cylinder valve is turned on.
    - Check to ensure that all hose connections are tight and free of leaks.
    - Check to ensure that the facepiece is in good condition and is clean.
    - Check to ensure that the harness system is in good condition and the straps are in the fully extended position.
    - Check to ensure that the bypass and mainline valves are operational. After checking the bypass valve, make sure that it is fully closed. (97-98)
89. The following are correct:
    - Follow the hoseline out if possible (male couplings away from the fire, female toward the fire).
    - Crawl in a straight line (hands flat on floor, move knee to hand).
    - Once in contact with the wall, crawl in one direction (all left-hand turns, all right-hand turns).
    - Call for directions, call out or make noise for other firefighters to assist you.
    - If possible, break a window or breach a wall to escape.
    - Activate your Personal Alert Safety System device. (102-103)

**3**

90. Earmuffs can compromise protection of the face by making it awkward to use SCBA and hoods; earplugs may melt when exposed to intense heat. (54)

91. Because cleanliness affects the performance of protective coats, trousers, and hoods. Clean outer shells have better fire resistance, and dirty turnouts absorb more heat. (62)

92. The respiratory tract can be damaged, and if the air is taken quickly enough into the lungs, there can be a serious decrease in blood pressure and failure of the circulatory system. Inhaling heated gases can cause pulmonary edema in the lungs, which can cause death from asphyxiation. (64)

93. A suspension of small particles of carbon, tar, and dust floating in a combination of heated gases (64)

94. Positive-pressure breathing apparatus maintains a slightly increased pressure in the user's facepiece, which helps prevent contaminants from entering the facepiece if a leak develops. (73)

95. That there is between 450 to 550 psi (3 100 kPa to 3 795 kPa) — depending on the manufacturer — of air remaining in the cylinder and that the firefighter and members of his or her SCBA team should leave the fire area *immediately* (76)

96. When the firefighter inhales, a partial vacuum is created in the regulator, causing the diaphragm to move inward and tilting the admission valve so that low-pressure air can flow into the facepiece. The diaphragm is held open until the firefighter exhales, and it moves back to its closed position. (74)

97. That they have different numbers of straps and that on some models, the regulator is attached to the facepiece rather than mounted on the harness (88)

98. By thoroughly washing with warm water containing any mild commercial disinfectant, rinsing with clear, warm water, and then drying with a lint-free cloth or allowing it to air dry (98)

99. Every three months (98)

100. Steel and aluminum cylinders, every five years; composite cylinders, every three years (99)

101. Put the cylinder into a shielded charging station, prevent cylinder overheating, be sure it is fully charged but not overpressurized, and under no circumstances place a composite bottle in water during a recharging operation. (99-100)

102. When the regulator becomes damaged or malfunctions. If this happens, the firefighter can close the mainline valve and open the bypass valve to provide a flow of air into the facepiece. (103)

103. Should the firefighter collapse or remain motionless for approximately 30 seconds, the PASS device will emit a loud, pulsating shriek. It can also be activated manually. (107)

# Chapter 4 Answers

## DEFINITIONS OF KEY TERMS

1. *Life safety rope* — Rope used to support rescuers and/or victims; can be only rope constructed of continuous filament fiber (115)
2. *Utility rope* — Rope used in any instance — excluding life safety applications — where the use of a rope is required (116)

## TRUE/FALSE

3. True (111)
4. False. The tensile strength of Type #1 manila rope is *much less than* that of nylon. (111)
5. True (115)
6. True (117)
7. False. The *bowline* knot is not a secure knot on synthetic fiber rope; therefore, it cannot be used in life safety situations. Use the figure of eight in place of the bowline. (117, 120)
8. True (123)
9. False. *Seventy-five* percent of the kernmantle rope's strength lies within its core. (124)

## MULTIPLE CHOICE

10. B (111)
11. A (112)
12. D (112)
13. C (120)
14. B (124)

## IDENTIFICATION

15. *NFPA 1983 — Standard on Fire Service Life Safety Rope, Harness, and Hardware* (115)
16. *Laid rope construction* — Rope that is constructed by twisting together yarns to form strands that are twisted together to make the final rope (114)
17. *Braided rope construction* — Rope that is constructed by uniformly intertwining strands of rope together and that has no core nor outer sheath (114)
18. *Braid-on-braid rope construction* — Rope that is constructed with both a braided core and a braided sheath (114)
19. *Kernmantle rope construction* — Rope constructed of main load-bearing strands (kern) covered with a braided sheath (mantle) (114-115)
20. A (112)

**4**

21. B, C (113)
22. A, B, C, E (112, 113, 114)
23. D (113)
24. B, C (113)
25. A (112)
26. B, C (113)
27. A, B, C, D (112, 113)
28. E (113)
29. Loop (116)
30. Round turn (116)
31. Bight (116)
32. Working end (117)
33. Standing part (117)
34. Running end (117)
35. A. *Knot* — Double figure of eight
    B. *Common use(s)* — Tying together ropes of equal diameters (120)
36. A. *Knot* — Becket or sheet bend
    B. *Common use(s)* — Joining two ropes; particularly suited for joining ropes of unequal diameters or joining a rope and a chain (121)
37. A. *Knot* — Clove hitch
    B. *Common use(s)* — Attaching a rope to an object such as a pole, post, or hose (118, 119)
38. A. *Knot* — Bowline
    B. *Common use(s)* — Forming a knot having a loop that will not constrict the object around which it is placed (117, 118)
39. A. *Knot* — Figure of eight on a bight
    B. *Common use(s)* — Attaching a rope to an anchor or tying-in a harness (120, 121)
40. *Ladder* — Use a bowline or figure of eight knot; slip it first through two of the rungs about one-third of the way down from the top, and pull the loop through. Then slip it over the top of the ladder. (128)
41. *Pike pole* — One method is to raise it head down having a clove hitch (with a safety) around the head and at least one half hitch, preferably two evenly spaced, around the handle. Another method is to raise it head up having a clove hitch toward the end of the handle and a half hitch in the middle of the handle and another around the head. (127-128)
42. *Axe* — Use a clove hitch, timber hitch, or girth hitch (each with a safety) around the head of the axe and at least one half hitch around the handle. (127)
43. *Smoke ejector* — Tie a bowline or figure of eight around two of the connecting rods between the front and back plates; also attach a tag line to the bottom of the unit. (128-129)
44. *Dry hoseline* — Fold the nozzle and hose end back over the rest of the hose forming a bight about 3 to 5 feet (1 m to 1.6 m) long, and tie the nozzle and hose together using a clove hitch with a half hitch safety knot. Then tie at least one half hitch around the bight. (128)
45. *Charged hoseline* — Tie a clove hitch with a half hitch safety knot around the hose about 1 foot (0.3 m) below the coupling and nozzle, and tie a half hitch through the nozzle bale so that it will hold the bale in the closed position as the hoseline is hoisted. (128)

**4**

46.  A.  Hand laundering
     B.  Using special rope washer
     C.  Using regular clothes washing machine (124)
47.  A.  By coiling it
     B.  By storing it in a canvas bag (125)

## SHORT ANSWER

48.  Its excellent resistance to mildew and rotting, excellent strength, and easy maintenance (112)
49.  Static rope stretches very little — 1½ to 2 percent — under normal loads, whereas dynamic lines stretch more than static lines both under weight and shock loads. Static line is most often preferred for rescue work; dynamic rope may be preferred under circumstances such as to arrest a fall or absorb the weight of a fall. (114)
50.  It eliminates the danger of the end of the rope slipping back through the knot and causing the knot to fail. (122)
51.  Core damage (123)
52.  Because 50 percent of the rope's strength is in the sheath (123)
53.  Visually inspect braided rope for cuts and fuzziness and for exterior damage, such as heat sears, caused by friction or fire; tactilely inspect it for permanent mushy spots. (123)
54.  Place slight tension on the rope and feel for lumps, depressions, or soft mushy spots in the rope; if the rope has any of these and is *undamaged,* the rope will return to its original texture in time. Carefully inspect the outer sheath as any damage indicates probable damage to the core. (123-124)
55.  Because water cannot be used in the cleaning process. It should be cleaned by wiping or gently brushing it to remove as much of the dirt and grit as possible. (124)
56.  By drying it in a dryer, by laying it on a flat, clean, dry surface out of direct sunlight, by hanging it in a hose tower, or by laying it on racks out of direct sunlight until it is completely dry (125)
57.  Placing the rope into a storage bag. This allows easy carrying of the rope and keeps dirt and grime from the rope. (126)
58.  The type of rope, its length, and its number (126)
59.  To prevent the object being raised or lowered from coming in contact with the structure or other objects (127)

# 5 | Chapter 5 Answers

## DEFINITIONS OF KEY TERMS

1. *Shoring* — The process of erecting a series of timbers or jacks to strengthen a wall or to prevent further collapse of a building or earth opening (158)
2. *Cribbing* — The process of arranging planks into a cratelike construction; this arrangement usually has separated joints (158)

## TRUE/FALSE

3. False. The combination spreader/shears powered hydraulic tool consists of two arms equipped with spreader tips that can be used for *pulling or pushing.* (134)
4. True (136)
5. False. There are *three* basic types of lifting bags: *high pressure, low and medium pressure, and leak-sealing.* (138)
6. False. The extremeties carry is a *two-person* carry that is *easy to perform with both conscious and unconscious victims.* (143)
7. True (143-144)
8. False. When moving a victim suspected of having a cervical spine injury, the rescuer *who applies and maintains axial traction* directs the other rescuers in their actions. (144)
9. False. When preparing to lower a victim from a window, the ladder should be raised to a point just *above* the window where the rescue is to be made. (147)
10. False. *Even* when a vehicle still has all of its wheels on the ground, stabilization of a vehicle involved in an accident is required to ensure maximum stability for extrication operations. (154)
11. True (157)
12. False. For cave rescues, it is recommended that rescuers carry *three* sources of light: a cap-mounted, battery-powered lantern, a waterproof flashlight, *and a candle.* (161)
13. True (161)
14. True (162)
15. True (162)
16. True (164)
17. False. Hoistway doors of an elevator are equipped with a weight or spring system that applies a constant force against the door toward the *closed* position. (167)
18. False. *Passengers can be instructed to* activate the elevator car's emergency stop switch. (168)

## MULTIPLE CHOICE

19. D (135)
20. B (134, 135)
21. B (138)
22. D (144)

23. C (152)
24. D (156)
25. A (157)
26. B (159)
27. B (162)
28. A (166)
29. C (167)

## MATCHING

30. C (146-147)
31. B (146)
32. A (146)

## IDENTIFICATION

33. *Extrication incidents* — Those incidents that involve the removal and treatment of victims who are trapped by some type of man-made machinery or equipment (133)
34. *Rescue incidents* — Those incidents that involve the removal and treatment of victims from situations involving natural elements, structural collapse, elevation differences, or any other situation not considered to be an extrication incident (133)
35. *Block* — Wooden or metal frame containing one or more pulleys called sheaves (139)
36. *Tackle* — Assembly of ropes and blocks through which the line passes to multiply the pulling force (139)
37. *Simple tackle* — One or more blocks reeved (threaded) with a single rope (140)
38. *Compound tackle* — Two or more blocks reeved with more than one rope (140)
39. *NFPA 1983 — Standard on Fire Service Life Safety Rope, Harness, and Hardware* (146)
40. *Primary search* — The first search of an area conducted in a quick, systematic fashion, checking first those areas where the percentage of finding victims is the highest (150)
41. *Secondary search* — The slower, more thorough and systematic search usually conducted after the fire has been controlled and the heat and smoke conditions within the building have improved somewhat to permit the more extensive search (150)
42. Running block (140)
43. Standing block (140)
44. Fall line (140)
45. Leading block (140)
46. *Lean-to collapse* — Collapse resulting when large sections of floor and roof sections drop, remaining in one piece and with support on one side and collapse on the other; a void underneath the supported side results. (157-158)
47. *Pancake collapse* — Collapse resulting when weakening or destruction of bearing walls cause floors or roof to collapse, causing the debris to fall as far as the lower floor or basement; voids between the layers of debris may result. (158)

**5**

48. *V-type collapse* — Collapse resulting when heavy loads are concentrated near the center of a floor, causing the floor to give way; voids near the walls result. (158)
49. A. Moon-shaped key — Inserted into the opening a few inches and then pulled downward (168)
    B. T-shaped key — Pushed straight into the opening to operate the locking mechanism (168)
    C. Drop key — Inserted into the opening far enough so that the front section of the key drops to form a 90-degree angle with the rear section and then rotated to unlock the door (168)

## LISTING

50. A. Spreaders
    B. Shears
    C. Combination spreader/shears
    D. Extension rams (133)
51. A. Vehicle-mounted air compressors
    B. Apparatus brake system compressors
    C. SCBA bottles
    D. Cascade system cylinders (137)
52. Answer should include at least eight of the following:
    - Plan the operation before starting the work.
    - Be thoroughly familiar with the equipment: its operating principles, methods, and limitations.
    - Consult individual operator's manuals and follow the recommendations for the specific system used.
    - Keep all components in good operating condition and all safety seals in place.
    - Have an adequate air supply and sufficient cribbing before beginning operations.
    - Position the bags on or against a solid surface.
    - Never inflate the bags against sharp objects.
    - Inflate the bags slowly and monitor them continually for any shifting.
    - Never work under a load supported only by bags.
    - Shore up the load with enough cribbing blocks to more than adequately support the load in case of bag failure.
    - Stop the procedure frequently to increase shoring or cribbing.
    - Be sure that the top layer is solid when using the box cribbing method; leaving a hole in the center may cause shifting and collapse.
    - Avoid exposing bags to materials hotter than 220°F (104°C). Insulate the bags with a nonflammable material. Remove a bag from service if any evidence of heat damage is noticed.
    - Never stack more than two bags; center the bags with the smaller bag on top, and inflate the botton bag first. (139)

5

53. Answer should include at least five of the following:
   - Be sure that the rope is the right size for the weight being lifted and the blocks being used.
   - When pulling on a fall line, everyone should exert a steady, simultaneous pull and hold onto the gain.
   - Be sure that the supports holding the standing and leading blocks will hold the load and the pull.
   - Pull in a direct line with the sheaves — not at an angle to either side.
   - Whenever possible, the pull should be downhill.
   - Pullers should stand so that they will be out of danger if the tackle or support fails.
   - Easing off on a suspended weight should be done gradually without jerking.
   - Hooks without safety latches should always be "moused" to prevent slings or ropes from slipping off. (140)
54. Answer should include at least eight of the following:
   - Wear full protective clothing and protective breathing apparatus when performing search and rescue operations in fire buildings.
   - Work in groups of two or more.
   - Attempt to locate more than one means of egress before entering the building.
   - Search on your hands and knees.
   - Search one room completely before moving to the next room.
   - Start the search on an outside wall, which allows you to ventilate by opening windows as soon as possible — provided positive-pressure ventilation is not to be used.
   - Move all furniture, searching behind and under each piece.
   - Search all closets and cupboards including shower stalls.
   - Pause occasionally during the search, and listen for cries for help or other audible signs or signals.
   - Move up and down stairs on your hands and knees; when ascending, proceed head first, and when descending, proceed feet first.
   - After searching a room, leave a sign or signs indicating that the room has been searched.
   - Look for extension of fire, and report any extension to the commanding officer.
   - Reach into the doorway or window with the handle of a tool if rooms or buildings are too hot to enter; frequently victims will be found just inside the door or windows.
   - Once you have successfully removed a conscious victim, place the victim in someone's custody so that he or she will not try to reenter the building for any reason. (150-151)
55. A. Through a normally operating door
   B. Through a window
   C. By compromising the body of the vehicle (153)
56. The following are correct:
   - How many and what type of vehicles are involved
   - Where the vehicles are positioned
   - Whether a fire is involved
   - Whether there are any hazardous materials involved
   - Utilities, such as gas or electricity, that may have been damaged and are posing a hazard to the victims and rescue personnel
   - Need for additional resources (153)

**5**

57. The following are correct:
- Number of victims in or around the vehicle and the severity of their injuries
- Condition of the vehicle
- Extrication tasks that may be required
- Any hazardous conditions that might exist
- Vehicles involved that may not be readily apparent
- Any victims who have been thrown clear of the vehicles
- Damage to structures or utilities that present a hazard
- Any other circumstances that warrant special attention (153)

58. A. Jacks
B. Air lifting bags
C. Cribbing (155)

59. A. The trench should be entered with proper protective equipment — this equipment includes head, hand, and eye protection. If a toxic or oxygen-deficient atmosphere is suspected, respiratory protection should also be worn.
B. Exit ladders should be placed in the trench on both sides and ends. Ladders should extend at least 3 feet (1 m) above the top of the trench and be secured in place if possible.
C. Firefighters should be careful with the tools they use in the trench to avoid injuring each other or the victim.
D. Unnecessary fire department personnel and bystanders should be kept out of the trench and away from its edge.
E. Rescuers should be aware of any other hazards that might exist at the scene such as underground electrical wiring, water lines, explosives, or toxic or flammable gases. (159-160)

60. A. Throw a rope to the victim.
B. Extend a long pole to the victim.
C. Throw a flotation device with an attached rope to the victim.
D. Use a boat to retrieve the victim.
E. Swim to the victim and drag the victim to safety. (163)

61. Step 1 — Survey the situation.
Step 2 — Seek expert assistance.
Step 3 — Neutralize power sources.
Step 4 — Stabilize machinery.
Step 5 — Survey again.
Step 6 — Extricate the victim.
Step 7 — Retrieve extrication equipment.
Step 8 — Secure the scene (164)

62. The following are correct:
- The medical condition and the degree of entrapment of the victim
- The number of rescue personnel required
- The type and amount of extrication equipment needed
- The need for special personnel, equipment, or expert assistance
- The level of fire or hazardous material hazard that is present (164)

63. The primary advantage is that the porta-power has accessories that allow it to be operated in narrow places in which the jack will not fit or cannot be operated. The primary disadvantage is that assembling complex combinations of accessories and actual operation of the tool are time-consuming. (136)

64. For most heavy lifting situations and as a compression jack for shoring or stabilizing operations (136)

65. For cutting through the roof, roof support columns or doorjambs, seat bolts, and door lock assemblies (137)

66. Because any sparks produced during cutting may provide an ignition source for flammable vapors (137)

67. To apply to cracks, open ends, or holes in low-pressure liquid storage containers or pipes (138)

68. After every use and periodically between use to make sure that they are in good condition (147)

69. To keep the victim away from the building as he or she is being lowered (148)

70. The firefighter with air remaining can share her or his air supply with the downed firefighter by using a buddy breathing connection — if SCBAs are so equipped. However, at no time should the firefighter remove his or her facepiece or compromise the proper operation of his or her SCBA in an attempt to share it with another firefighter or victim. (152)

71. Providing additional support to key places between the vehicle and the ground or other solid anchor points in order to prevent any further movement of the vehicle (154)

72. Because doing so simplifies the framing required to prevent a cave-in (159)

73. The rescuer's first priority must be to uncover enough of the victim's head and chest to allow for sufficient respirations; an air hose or partly opened cylinder can be inserted into a hole dug to the victim's face; and in an emergency, air may be directed to a victim through a garden hose. (159)

74. By pushing themselves across the ice in a flat-bottom boat or raft or by operating from ladders or a sheet of plywood laid on the ice to distribute their weight (164)

# 6 | Chapter 6 Answers

## DEFINITIONS OF KEY TERMS

1. *Doorjamb* — The side of a doorway opening (183)
2. *Rabbeted jamb* — A doorjamb into which a shoulder has been milled to permit the door to close against the provided shoulder (183)
3. *Stopped jamb* — A doorjamb inside of which a wooden strip or doorstop is attached against which the door closes (183-184)
4. *Tempered plate glass* — Glass that has been heat tempered, resulting in high tension stresses in the center of the glass and high compression stresses on the exterior surfaces; is approximately four times stronger than regular plate glass (190)
5. *Lexan® plastic* — A polycarbonate that is widely used as a glass substitute because of its ability to withstand abuse from vandalism or weather (198)

## TRUE/FALSE

6. False. Generally, *power saws* are used for forcible entry, and *handsaws* are used for rescue. (174)
7. True (175)
8. True (178)
9. True (180)
10. False. Residential swinging doors generally open *inward*, and those in public buildings should open *outward*. (183)
11. True (184)
12. True (194)
13. False. *Automatic closing* fire doors normally remain open but close when heat actuates the closing device. (194)
14. False. The most effective entry through Lexan is by *using a circular saw equipped with a carbide-tipped blade*. (198)
15. False. Generally, the floors of upper stories of family dwellings are *wood joist with subfloor and finish construction*. (199)
16. False. When cutting open a metal wall, the firefighter should cut *along the studding, which will provide stability for the saw and allow ease of repair*. (201)

## MULTIPLE CHOICE

17. A (173)
18. C (173-174)
19. D (176)
20. A (178)
21. D (183)
22. B (184)
23. C (185)
24. D (186)

25. A (190)
26. C (192)
27. B (192)
28. B (197)
29. B (199)
30. A (200)
31. D (201)

## IDENTIFICATION

32. *Hollow core door* — Slab door with a core consisting of an assembly of wood strips formed into a grid or mesh and glued within a frame (183)

33. *Solid core door* — Slab door with a core that is constructed of solid material, such as tongue-and-groove blocks or boards glued within a frame or a compressed mineral substance that is fire resistant (183)

34. *Rapid entry key box* — Box that contains all necessary keys to a building, storage areas, gates, and elevators; that is mounted at a high-visibility location on the building's exterior; and that can be opened by fire department personnel (194)

35. *Class A openings* — Openings located in walls separating buildings or in walls within a building that is separated into distinct fire areas (194)

36. *Class B openings* — Vertical enclosures, such as elevators, stairways, or dumbwaiters, that may allow fire to spread within a building (194)

37. *Penetrating nozzle* — Special-purpose nozzle designed to penetrate masonry and some concrete (200)

38. CU (174)

39. PR (178)

40. PU (179)

41. ST (180)

42. CU, ST (173-174)

43. PU (178)

44. ST (180)

45. PR (178)

46. PR (178)

47. CU (174)

48. PR (178)

49. CU (176)

50. ST (180)

51. CU (176)

52. PU (178-179)

53. ST (180)

54. PR (178)

55. *Wooden handles* — Check for cracks, blisters, or splinters; sand to minimize hand injuries; clean with soapy water, rinse, and dry after use; apply a coat of boiled linseed oil to prevent roughness and warping; check to ensure that the head is on tight. (180-181)

56. *Fiberglass handles* — Wash with warm, soapy water; dry with a soft, dry cloth; check to ensure that the head is on tight. (181)

57. *Cutting edges* — Check to ensure that the cutting edge is free of nicks or tears; replace cutting edge of bolt cutters when needed; file the edges by hand — grinding takes the temper out of the metal. (181)

58. *Plated surfaces* — Inspect for damage; wipe plated surfaces clean or wash with soap and water; do not paint axe heads. (181)

59. *Unprotected metal surfaces* — Keep clean of rust; keep oiled when not used; do not completely paint; check to see that the metal surfaces are free of burred or sharp edges — file off any found. (181)

**6**

60. *Power equipment* — Check to see whether the equipment starts manually; check blades and equipment for completeness and readiness; check electric tool cords for cuts and frays; make sure that the appropriate guards are in place. (181)
61. *Panic-proof type* — Push or press the doors or wings in opposite directions. (191)
62. *Drop-arm type* — On the door to which the arm of the mechanism passes, press the pawl to disengage it from the arm, and then push the wing to one side. (191)
63. *Metal-braced type* — Unhook an arm that holds one of the doors in place, and fasten it back against the fixed door. (191)
64. *Wooden checkrail* — If the sashes are locked at the center of the checkrail, pry at the center of the lower sash. (196)
65. *Metal checkrail* — Because sash locks are not likely to give under prying, break the glass near the lock and unlock the window from the inside. (196)
66. *Casement* — Break the lowest glass pane, force or cut the screen in the area, unlock the latch, operate the crank or lever to open the window, and completely remove the screen. (196-197)
67. *Projected* — Remove the screen either before or after breaking the glass (depending upon whether the window projects in or out), break the lowest glass pane, unlock the latch, and open the window. (197)
68. *Awning or jalousie* — Avoid forcing entry through this type, but if necessary, remove several panels of glass to allow enough room for entrance. (197)

## LISTING

69. A.  Match the saw to the task and the material to be cut. Never push a saw beyond its design limitations.
    B.  Always wear proper protective equipment, including gloves and eye protection.
    C.  Do not use any power saw when working in a flammable atmosphere or near flammable liquids.
    D.  Keep unprotected and nonessential people out of the work area.
    E.  Follow manufacturer's guidelines for proper saw operation.
    F.  Keep blades and chains well sharpened. A dull saw is more likely to cause an accident than a sharp one. (175-176)
70. The following are correct:
    - Store and use acetylene cylinders in an upright position to prevent loss of acetone.
    - Handle cylinders carefully to prevent damage to the cylinder and to the filler.
    - Avoid exposing cylinders to excessive heat (ambient air temperature exceeding 130°F [54°C]).
    - Do not place cylinders on wet or damp surfaces.
    - Store acetylene cylinders in an area separate from oxygen cylinders and other oxidizing gas cylinders.
    - Perform a soap test (applying a solution of soap and water on fittings) to detect leaks at regulator, torch, hose, and cylinder connections.
    - Open acetylene cylinder valves no more than a three-quarter turn. Do not use wrenches on cylinders that have handle valves.

- Do not use acetylene at pressures greater than 15 psi (103 kPa).
- Do not exceed the withdrawal rate of one-seventh of the cylinder capacity per hour.
- Keep valves closed when cylinders are not in use and when they are empty. (177-178)
71. A. What type of door it is
    B. How the door is hung
    C. How it is locked (184-185)
72. A. Feeling the wall for hot spots
    B. Looking for discolored wallpaper or blistered paint
    C. Listening for the sound of burning
    D. Using electronic sensors (202)

## SHORT ANSWER

73. Tool failure (including breakage), injury to the operator (174)
74. Because the saw's motor or sparks from the cutting operation can ignite, causing a fire or explosion (174)
75. That a person can generate more force to an object using such a tool than they could without one (178)
76. If the blade is extremely sharp and its body is ground too thin, pieces of the blade may be broken out when cutting roofs or striking nails and other materials in flooring; if the body of the blade is too thick, it is difficult to drive the axe head through ordinary objects. (180)
77. Wear full protecting clothing — and if making entry into a fire building, wear SCBA and have charged hoselines in place — stand to windward side of the glass pane, keep hands above the point of impact, and strike the tool at the top of the pane. After the glass is broken, remove all jagged pieces from the sash. (185-186)
78. Near the lock, insert the tool blade between the door and jamb, forcing the blade in and against the rabbet or stop. Pry the tool bar away from the door to move the door and jamb apart. When the lock has cleared the keeper, pull the door open or pry it open with another tool. (187-188)
79. If the door has a stopped jamb, the stop may be removed completely before inserting the tool between the door and the jamb near the lock; then the door may be moved enough to permit the bolt to pass the keeper. If the door has a rabbeted jamb, it may be necessary to use two tools to pry the door sufficiently to allow the bolt to pass the keeper. (188)
80. They may be forced open by inserting a wedge tool between the jamb and door near the lock and prying the door away from the frame. "Burglar blocks" — metal rods or special devices — may sometimes bar or block patio sliding doors, practically eliminating any possibility of forcing without causing excessive damage. (191-192)
81. The firefighters should block fire doors open to prevent their closing and trapping them. (195)
82. By using a compressed-air or electric jackhammer (200)
83. Remove the siding, sound the wall for stud supports, and cut the drywall or plaster along a stud; break through the interior side of the wall. (202)

# 7 | Chapter 7 Answers

1.  *Ventilation* — The systematic removal and replacement of heated air, smoke, and gases from a structure with cooler air (207)
2.  *Mushrooming* — The lateral spreading and banking down of heat, smoke, and fire gases that have accumulated and are trapped by the roof or ceiling (210)
3.  *Flashover* — The condition in which an entire room becomes involved in flames as a result of excessive heat causing all the combustibles in the room to reach their ignition temperatures (210)
4.  *Backdraft* — The sudden ignition, caused by the admittance of a previously lacking air supply, of combustible materials that have reached their ignition temperatures but have not already ignited because of the lack of oxygen to support combustion (210-211)
5.  *Vertical ventilation* — The opening of the roof or existing roof openings for the purpose of allowing heated gases and smoke to escape to the atmosphere (218)
6.  *Stack effect* — The natural, vertical movement of heat and smoke throughout a high-rise building (214)
7.  *Horizontal ventilation* — The venting of heat, smoke, and gases through wall openings such as windows and doors (230)

**TRUE/FALSE**

8.  True (212)
9.  False. Slate and tile roofs *can* be opened by using a *large sledgehammer to smash the slate or tile and the boards underneath.* (225-226)
10. True (226)
11. True (226)
12. True (228)
13. False. For hydraulic ventilation, the fog stream should be set on a wide fog pattern that will cover *85 to 90* percent of the window or door opening from which the smoke will be pushed. (236)
14. True (237)

**MULTIPLE CHOICE**

15. C (222)
16. C (223)
17. C (226)
18. B (226-227)
19. D (230)

## IDENTIFICATION

20. *Trench ventilation* — Ventilation used to stop the spread of fire in a long, narrow structure and performed by cutting a large hole, or trench, that is at least 4 feet (1.3 m) wide and extends from one exterior wall to the opposite exterior wall (227)

21. *Strip ventilation* — Ventilation used to stop the spread of fire in large structures and performed by cutting open large strips in the roof, leaving large holes that do not extend completely across the building (228)

22. *Mechanical ventilation* — Forced ventilation accomplished with blowers or ejectors (232)

23. *Hydraulic ventilation* — Forced ventilation accomplished with fog streams (232)

24. Butterfly (219)

25. Lantern (219)

26. Hip (219)

27. Mansard (219)

28. Shed (219)

29. Gambrel (219)

30. A. *Flat* — Flat roofs are usually constructed with wooden, concrete, or metal joists covered with sheathing, which is covered with a layer of waterproofing material and an insulating material; such roofs may be constructed of reinforced concrete, precast gypsum, or concrete slabs set within metal joists. They may or may not have a slight slope to facilitate water drainage; are frequently pierced by chimneys, vent pipes, shafts, scuttles, and skylights; may be surrounded and/or divided by parapets; and may support water tanks, air-conditioning equipment, and other objects. (222)

    B. *Pitched* — Pitched roofs are constructed with timber rafters or metal trusses that run from the ridge to a wall plate on top of the outer wall at the eaves level. These rafters or trusses are covered with sheathing boards, roofing paper, possibly roll felt with asphalt roofing tar, and shingles (wood, metal, composition, asbestos, slate, or tile). They are elevated in the center, sloping to the edges and have a more pronounced downward incline than do flat roofs. (224)

    C. *Arched* — They may be constructed with bow-string trusses for supporting members, and a ceiling may be constructed, concealing the trusses and forming a cockloft; there are also trussless arched roofs made of a network of short timbers of uniform length that are bolted together at angles. Trussless arch construction enables all parts of the roof to be visible. (226)

31. A. Windward (230, 231)
    B. Leeward (230, 231)

## LISTING

32. The following are correct:
    - Proper ventilation simplifies and expedites rescue by removing smoke and gases that endanger trapped or unconscious occupants and improves visibility so that unconscious victims may be located easier. (208)
    - Proper ventilation can localize the fire, and it permits firefighters to more rapidly locate the fire and proceed with extinguishment. (208-209)

**7**

- By contributing to the fire's rapid extinguishment, ventilation assists in reducing water, heat, and smoke damage. (209)
- Proper ventilation of a building during a fire reduces the possibility of mushrooming. (210)
- Ventilation helps to alleviate the conditions leading to flashover because the heat is removed before it reaches the necessary levels for mass ignition. (210)
- Top ventilation reduces the potential of a backdraft occurring. (210-211)

33. A. Is there a need for ventilation at this time?
    B. Where is ventilation needed?
    C. What type of ventilation should be used? (212)

34. The following are correct:
    - Building type and design
    - The number and size of wall openings
    - The number of stories, staircases, shafts, dumbwaiters, ducts, and roof openings
    - The availability and involvement of exterior fire escapes and exposures (212-213)

35. Answers should include at least eight of the following:
    - Observe the wind direction with relation to exposures.
    - Work with the wind at your back or side to provide protection while cutting the roof opening.
    - Note the existence of obstructions or excess weight on the roof.
    - Provide a secondary means of escape for crews on the roof.
    - Exercise care in making the opening so that main structural supports are not cut.
    - Guard the opening to prevent personnel from falling into the building.
    - Evacuate the roof when ventilation work is complete.
    - Use lifelines, roof ladders, or other means to protect personnel from sliding and falling off the roof.
    - Exercise caution in working around electric wires and guy wires.
    - Ensure that all personnel on the roof are wearing full personal protective equipment, including SCBA.
    - Keep other firefighters out of range of the axe.
    - Caution axe users to beware of overhead obstructions within the range of their axe.
    - Start power tools on the ground to ensure operation; however, it is important that the tool be shut off before hoisting or carrying the tool to the roof.
    - Make sure that the angle of the cut is not toward your body.
    - Extend ladders at least three rungs above the roof line. When possible, 6 feet (2 m) or more is desirable. When using elevating platforms, the floor of the platform should be even with or slightly above roof level.
    - "Sound" the roof for structural integrity before stepping onto it; do not jump without checking it first.
    - Use supporting members of the structure for travel; use no diagonal travel.
    - Use pre-incident planning and inspections to identify buildings that have roofs supported by lightweight or wooden trusses. Realize that these roofs may fail early into a fire and are extremely dangerous to be operating on or below.
    - When using a roof ladder, make sure that it is firmly secured over the peak of the roof before operating from it.

- Be aware of warning signs of an unsafe roof condition.
- Work in groups of at least two, with no more people than absolutely necessary to get the job done. (219-221)

36. The following are correct:
    - Melting asphalt
    - "Spongy" roof — a normally solid roof that springs back when walked upon
    - Smoke coming from the roof
    - Fire coming from the roof (221)

37. The following are correct:
    - Improper use of forced ventilation
    - Excess breakage of glass
    - Fire streams directed into ventilation holes
    - Breakage of skylights
    - Explosions
    - A burn-through
    - Additional openings between the attack team and the upper opening (229)

38. The following are correct:
    - The opening of a door or window on the windward side of the structure prior to first opening a door on the leeward side
    - The opening of doors and windows between the advancing fire fighting crews and the established ventilation exit point
    - The interruption of the established current caused by a firefighter or other obstruction in the doorway (231-232)

39. The following are correct:
    *Advantages*
    - It ensures more positive control.
    - It supplements natural ventilation.
    - It speeds the removal of contaminants, facilitating a more rapid rescue under safer conditions.
    - It reduces smoke damage.
    - It promotes good public relations. (232-233)
    *Disadvantages*
    - The introduction of air in such great volumes can cause the fire to intensify and spread.
    - It is dependent upon a power source.
    - It requires special equipment. (233)

40. Answer should include at least four of the following:
    - Firefighters can set up forced ventilation procedures without entering the smoke-filled environment.
    - Because positive-pressure ventilation merely supplements natural ventilation currents, it is equally effective with horizontal or vertical ventilation.
    - It allows more efficient removal of smoke and heat from the structure or vessel.
    - The velocity of air currents within the building are minimal and have little, if any, effects that disturb the building contents or smoldering debris. Yet, the total exchange of air within the building is faster than using negative-pressure ventilation.
    - The placement of blowers does not interfere with ingress or egress.

**7**

- The cleaning and maintenance of blowers is greatly reduced compared to that of smoke ejectors.
- This system is applicable to all types of structures or vessels and is particularly effective at removing smoke from large, high-ceiling areas where negative-pressure ventilation is ineffective.
- Heat and smoke may be directed away from unburned areas or paths of exit. (234)

## SHORT ANSWER

41. To ensure that it cannot close and allow the shaft to become filled with superheated gases after ventilation tactics are started (215)

42. It may cause the fire to spread throughout areas of the building that otherwise would not have been affected. (216)

43. It is typically quicker to open an existing opening than to cut a hole in the roof. However, existing openings are rarely large enough or in the best location, and they usually supplement holes that have to be cut. (222)

44. Determine the location of the opening. Using a roof ladder and working upwind from the opening location, sound for solid support or rafters, marking their locations. Remove roof covering. Cut decking or sheathing alongside a joist or rafter, and then cut opposite side of the opening in a like manner. Pry up decking or sheathing boards and remove. Push through ceiling with the blunt end of a pike pole or other long-handled tool. (223, 224-225)

45. The orderly movement of fire gases from the building will either be upset or destroyed, forcing superheated air and gases back down on firefighters, causing serious injury or death. Such streams also will contribute to the spread of fire throughout the structure. (229)

46. Negative-pressure ventilation is mechanical forced ventilation in which smoke ejectors are used to develop artificial circulation and pull smoke out of a structure. Smoke ejectors are placed in windows, doors, or roof vent holes and pull smoke, heat, and gases from inside the building and eject them to the exterior. The ejectors should be placed to exhaust in the same direction as the natural wind. (233)

47. Positive-pressure ventilation is a forced ventilation technique that uses the principle of creating pressure differentials. High-volume blowers placed several feet from an exterior doorway are used to blow fresh air into the structure. The smoke is then ejected from another opening on the opposite side of the structure. It is important that no other exterior openings be opened during such ventilation. (234-235)

# Chapter 8 Answers

## DEFINITIONS OF KEY TERMS

1. *Base section* — The bottom section of an extension ladder; also called bed section or main section (241)
2. *Beam* — The side rail of a ladder (241)
3. *Beam bolts* — Bolts that pass through both rails at the truss block of a wooden ladder to tie the two truss rails together (241)
4. *Butt* — The bottom end of the ladder that will be placed on the ground or other supporting surface when the ladder is raised; also called heel (241)
5. *Butt spurs* — Metal safety plates or spikes attached to the butt of ground ladder beams (241)
6. *Fly* — The upper section or top sections of an extension ladder or aerial device (242)
7. *Guides* — Wood or metal strips, sometimes in the form of slots or channels, on an extension ladder that guide the fly section while being raised (242)
8. *Halyard* — A rope or cable used for hoisting and lowering the fly sections of a ground ladder (242)
9. *Heat sensor label* — A label affixed to the ladder beam near the tip to provide a warning that the ladder has been subjected to excessive heat (242)
10. *Hooks* — A pair of sharp curved devices at the top of a roof ladder that fold outward from each beam (242)
11. *Pawls* — Devices attached to the inside of the beams on fly sections used to hold the fly section in place after it has been extended; also called dogs or locks (242)
12. *Protection plates* — Plates fastened to a ladder to prevent wear at points where it comes in contact with mounting brackets (242)
13. *Pulley* — A small, grooved wheel through which the halyard is drawn on an extension ladder (242)
14. *Rails* — The two lengthwise members of a trussed ladder beam that are separated by truss or separation blocks (242)
15. *Rungs* — Cross members (usually round or oval) between the beams on which the climber steps (242)
16. *Safety shoes* — Rubber or neoprene foot plates, usually of the swivel type, attached to the butt of the beams of a ground ladder (242)
17. *Spurs* — Metal points at the lower end of staypoles (242)
18. *Staypoles* — The poles attached to long extension ladders to assist in raising and steadying the ladder; may be permanently attached or removable; also called tormentor poles (243)
19. *Stops* — Wood or metal pieces that prevent the fly section from being extended too far (243)
20. *Tie rods* — Metal rods running from one beam to the other (243)
21. *Toggle* — A hinge device by which a staypole is attached to a ladder (243)
22. *Top or tip* — The extreme top of a ladder (243)
23. *Truss block* — Separation pieces between the rails of a trussed ladder; sometimes used to support rungs (243)

**8**

24. True (247)
25. False. One difference in telescoping aerial platform apparatus and aerial ladder platform apparatus is that an *aerial ladder platform apparatus* is designed with a large ladder that allows firefighters to routinely climb back and forth from the platform. (247)
26. True (247-248)
27. True (249-250)
28. True (262)
29. False. A residential story will average *8 to 10 feet (2 m to 3 m)* from floor to floor, with a 3-foot (1 m) distance from the floor to the windowsill; stories of commercial buildings will average *12 feet (4 m)* from floor to floor, with a 4-foot (1.3 m) distance from the floor to the windowsill. (258)
30. True (271)
31. False. Whenever possible, a ladder should be extended *after* it is pivoted. (273)
32. False. Typically ladders of 35 feet (11 m) or larger should be raised by at least *three* firefighters. (280)

MULTIPLE CHOICE

33. A (249)
34. C (252-254, 256-257)
35. B (268)
36. D (268)
37. D (269)
38. C (272)
39. C (272-273)
40. A (280)

IDENTIFICATION

41. *NFPA 1931 — Standard on Design of and Design Verification Tests for Fire Department Ground Ladders* (241)
42. *NFPA 1932 — Standard on Use, Maintenance, and Service Testing of Fire Department Ground Ladders* (244)
43. *NFPA 1901 — Standard for Pumper Fire Apparatus (259)*
44. *NFPA 1904 — Standard for Aerial Ladder and Elevating Platform Fire Apparatus* (246)
45. *Aerial apparatus* — A fire apparatus equipped with a powered extension ladder, telescoping aerial platform, or articulating aerial platform (241)
46. *Ground ladder* — A ladder that is manually carried to the desired position and manually raised, positioned, and lowered (242)
47. *Single ladder* — A ladder consisting of one section; also called straight ladder (242)
48. *Roof ladder* — A single ladder equipped with folding hooks at the top which provide a means of anchoring the ladder over the roof ridge or other roof part (244)
49. *Folding ladder* — A single ladder having hinged rungs allowing the ladder to be folded so that one beam rests against the other (244)
50. *Extension ladder* — A ladder consisting of a base section and one or more fly sections that travel in guides or brackets to permit length adjustment (242, 244)

5

## DEFINITIONS OF KEY TERMS

## MULTIPLE CHOICE

| | | |
|---|---|---|
| 21. D (402) | 28. B (411) | 35. B (419) |
| 22. C (403) | 29. D (412) | 36. C (420) |
| 23. B (403) | 30. C (413) | 37. B (421) |
| 24. B (404) | 31. C (415) | 38. D (425) |
| 25. B (406) | 32. C (416) | 39. D (429) |
| 26. C (409) | 33. A (418) | 40. C (430) |
| 27. B (410) | 34. C (419) | 41. C (430) |

## IDENTIFICATION

42. *Company officer* — The leader of members of a fire department company; responsible for making tactical deployment decisions while assisting and supervising individual company members (401)

43. *Apparatus driver/operator* — The driver/operator of fire apparatus; responsible for the safe operation of the apparatus and all associated equipment and for the safe transportation of the crew to and from the fire scene (401)

44. *"O" pattern* — Method of water application used in a combination attack in which the stream is directed at the ceiling and rotated so that the stream edge reaches the ceiling, wall, floor, and opposite wall (405)

45. *LPG* — Liquefied petroleum gas; a fuel gas stored in a liquid state under pressure (411)

46. *SOPs* — Standard operating procedures; a fire department's predetermined plans or written policies (417)

47. *O-A-T-H method* — Method of signaling (by tugging a line) used by a firefighter inside a building and the safety person outside to communicate with each other; one tug — O for "OK"; two tugs — A for "Advance"; three tugs — T for "Take up"; four tugs — H for "Help" (426)

48. *Perimeter* — The boundary of a wildland fire; the total length of the outside edge of the burning or burned area (430)

49. Both (410, 411)
50. Natural gas (410)
51. Both (410, 411)
52. Both (410, 411)
53. Natural gas (410)
54. Both (410, 411)
55. Natural gas (410)
56. LPG (411)
57. LPG (411)
58. LPG (411)
59. LPG (411)
60. Spot fire (429)
61. Head (429)
62. Finger (429)
63. Right flank (429)
64. Rear (429)
65. Left flank (429)

66. A. *Direct attack* — Directing water at the base of the fire from a position close to the fire and using a solid stream or straight stream (404)

    B. *Indirect attack* — Directing water at the ceiling and back and forth at the superheated gases at the ceiling level and using either a solid, straight, or narrow fog pattern (405)

    C. *Combination attack* — Directing water around the room, both at the base of the fire and at the ceiling level, beginning by directing the stream at the ceiling and then near the floor and using either a solid, straight, or penetrating fog stream (405)

67. A. By using high-expansion foam to flood the basement and extinguish the fire (423)

    B. By using cellar nozzles put through holes in the floor above the fire to provide water to the fire (423)

68. A. *Ground fuels (duff)* — Small twigs, leaves, and needles that are decomposing on the ground (427)

    B. *Surface fuels* — Living surface vegetation including grass, brush, and other low vegetation; nonliving surface vegetation including downed logs, heavy limbs, etc. (428)

    C. *Crown fuels* — Suspended and upright fuels physically separated from the ground fuels to the extent that air can circulate freely around the fuels causing them to burn more readily (428)

## LISTING

69. The following are correct:
    - Imminent building collapse
    - Fire that is behind, below, or above the attack team
    - Kinks or obstructions to the hoseline
    - Holes, weak stairs, or other fall hazards
    - Suspended loads on fire-weakened supports
    - Hazardous or highly flammable commodities likely to spill
    - Backdraft or flashover behavior
    - Electrical shock hazards
    - Overexertion, confusion, or panic by team members (405)

70. The following are correct:
    - Fire conditions
    - Volume of water needed for extinguishment
    - Reach needed
    - Number of persons available to handle hoseline
    - Mobility requirements
    - Tactical requirements
    - Speed of deployment
    - Potential fire spread (406)

**12**

71. A. As a cooling agent to extinguish fires and to protect exposures (408)
    B. As a mechanical tool to move fuel, whether or not it is burning (409)
    C. As a substitute medium to displace fuel from pipes or tanks that are leaking (409)
    D. As a protective cover for advancing teams to shut off liquid or gaseous fuel valves (409)
72. Answer should include at least five of the following:
    - Increased life safety risks to firefighters from traffic
    - Increased life safety risk to passing motorists
    - Reduced water supply
    - Difficulty in determining the products involved
    - Difficulty in containing spills and runoff
    - Tanks and piping weakened or damaged by the force of collisions
    - Instability of vehicles
    - Additional concerns (residential neighborhood, schools, etc.) posed by the surroundings of the incident (409-410)
73. A. The path of electricity through the body
    B. The degree of skin resistance
    C. Length of exposure
    D. Available current
    E. Available voltage
    F. Current frequency (412)
74. A. Rescue
    B. Fire control
    C. Property conservation (417)
75. The following are correct:
    - Placement of lines to intervene between trapped occupants and fire or to protect rescuers
    - Protection of primary means of egress
    - Protection of interior exposure
    - Protection of exterior exposure
    - Initiation of extinguishment from the unburned side
    - Operation of master streams (418-419)
76. The following are correct:
    - Back up the initial attack line.
    - Protect secondary means of egress.
    - Prevent fire extension.
    - Protect the most severe exposure.
    - Assist in extinguishment. (419)
77. A. Fuel
    B. Weather
    C. Topography (427)

**12**

## SHORT ANSWER

78. A portable light, an axe, and a prying tool of some type (402)
79. Small, exterior fires such as trash dumpsters and small brush fires (406)
80. Because the fuel can be absorbed within their protective clothing in a "wicking" action, which can result in contact burns of the skin and flaming clothing if an ignition source is present (407)
81. Firefighters should not extinguish the fire unless the leaking product can be shut off and should try to contain the pooling liquid until the flow can be stopped, control all ignition sources in the leak area, and listen for an increase in the intensity of sound or fire issuing from a relief valve, which may indicate that rupture is imminent. (408)
82. When flames contact the vapor space of the flammable liquid vessel and insufficient water is applied to keep the tank cool (408)
83. Bills of lading, manifests, placards, or the driver of the transport vehicle; in cases where these cannot be found, from the shipper or manufacturer responsible for the vehicle (410)
84. No. If the fire is extinguished, the vapors could build up and then reignite with disastrous results. (411, 412)
85. The flow of electricity to the object involved should be stopped. (413)
86. Because the hazard of shock is greater and because extensive damage may occur to electrical equipment not involved in the fire (414)
87. To coordinate the overall activities at the scene by constantly evaluating the allocation of resources and the need for additional resources and by coordinating with other entities such as mutual aid units, EMS personnel, utility crews, and members of the media (421)
88. To first extinguish any ground fire around or under the vehicle and then to attack the remaining fire in the vehicle (424)
89. The possibility that they could become trapped between the two fires (430)

## CASE STUDIES

*Case Study 1*

The following are suggested actions that should be taken by the companies. Your local procedures may differ.
A. *Engine 5* — After the supply line has been laid into the scene, the firefighter at the hydrant charges the line once the hose clamp is in place at the rear of the apparatus. Because of the rescue and fire situation, it would probably be wise for the captain to pass command to the next arriving unit. Pull and advance a hoseline of at least 1½-inch (38 mm) diameter in order to cut the fire off in the room of origin and to prevent it from entering the right rear bedroom where the victim is supposedly located. If possible, search the bedroom that is supposed to contain the victim.
B. *DC 602* — Assume the command, as it was passed by Engine 5. Evaluate conditions, report them to dispatch, and call for more resources, if required. Make assignments for incoming units. Send the driver around the back of the structure to report on conditions from that side.

C. *Engine 11* — Stop at the hydrant with the Humat valve on it, and check with the driver/operator of Engine 5 to see whether Engine 11 will need to connect to the valve and boost the pressure on the original supply line. Once the water supply is established, personnel from Engine 11 should pull another handline from Engine 5 and provide backup for the first hose team. Also make sure exposures are protected.

D. *Truck 5* — Once on the scene, the truck company members should make sure that appropriate ventilation is effected, and members should assist the engine companies with search and rescue and the removal of the victim. Salvage and overhaul may be done after the fire is controlled.

E. *Rescue Squad 18* — While no guidelines for these units are given in the **Essentials** manual, the following duties are some that may be considered:
- Assist other crews, particularly the truck company, with their assigned duties.
- Administer emergency medical treatment to victims.
- Perform auxiliary functions such as scene lighting, SCBA refilling, etc.

*Case Study 2*

The following are suggested answers.

A. The car is already lost; the brush fire is the primary concern. Based on the time of year, the firefighters must be concerned with the cornstalks being dead, dry, and very subject to rapid fire spread. The hay barn is also probably full of hay or straw. Should the barn catch fire, it will be difficult to protect the other exposures.

B. Figure 12-B shows the possible placement of Engine and Tanker 33.

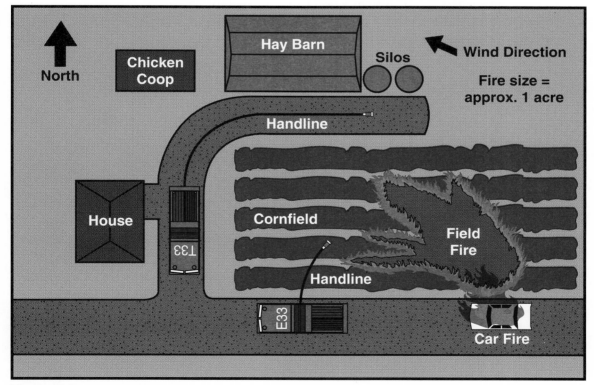

**Figure 12-B**

Engine 33 should be put in a position to make an attack on the field fire. This should only be done if safely possible. Tanker 33 should be backed into the driveway to protect the exposed structures. Booster lines or larger lines may be used depending on conditions.

**12**

C.  Brush Pumpers 33 and 20 should attack the field fire. In this case, a direct attack from the burned area would be the best plan of action. The fire is small and controllable. Also, there is little room between the exposures and the main fire for a backfire to be lit. Chances are that brands from the backfire could blow into the barn and escalate the incident. Once the field fire is controlled, any of the units can complete extinguishment of the car fire.

**13**

# Chapter 13 Answers

## DEFINITIONS OF KEY TERMS

1.  *Indicating valve* — A valve whose position as open or closed can be determined at a glance (439)
2.  *Accelerator* — Quick-opening device that unbalances the differential in the dry-pipe valve causing it to trip more quickly (445)
3.  *Exhauster* — Quick-opening device that quickly expels the air from the dry-pipe system (445)
4.  *Retarding device* — A part of the alarm check valve that catches excess water that may be sent through the alarm valve during momentary water pressure surges in order to reduce the chance of a false alarm activation (444)

## TRUE/FALSE

5.  True (434)
6.  True (435)
7.  False. The three basic sprinkler designs are upright, pendant, and sidewall; *the upright standard sprinkler cannot be inverted for use in the hanging or pendant position.* (437)
8.  False. The main control valve of the sprinkler system should always be returned to the *open* position after maintenance is complete. (438)
9.  False. Main water control valves are *indicating* valves and are *manually* operated. (439)
10.  False. The water supply for sprinkler systems is designed to supply *only a fraction* of the sprinklers actually installed on the system. (441)
11.  True (446)

## MULTIPLE CHOICE

12.  C (434)
13.  B (442)
14.  B (444)
15.  A (446)

16.  B (448)
17.  C (448)
18.  C (449)

19. *NFPA 13 — Standard for the Installation of Sprinkler Systems* (433)
20. *NFPA 13D — Standard for the Installation of Sprinkler Systems in One- and Two-Family Dwellings and Mobile Homes* (433)
21. *OS&Y valve* — Outside screw and yoke valve; valve having a yoke on the outside with a threaded stem that controls the opening and closing of the valve gate (439)
22. *PIV* — Post indicator valve; valve having a hollow metal post attached to the valve housing inside which is a valve stem with an attached target with the words "open" and "shut" to indicate valve position (439)
23. *WPIV* — Wall post indicator valve; post indicator valve that extends through the wall with the target and valve operating nut on the outside of the building (439)
24. *PIVA* — Post indicator valve assembly; post indicator valve that uses a butterfly valve rather than a gate valve (439)
25. *Fire department connection* — Connection that allows additional water and pressure to be supplied by a pumper to a sprinkler system; consists of a siamese with at least two 2½-inch (65 mm) female connections or one large-diameter connection that is connected to a clappered inlet (441)
26. *Ball drip valve* — Valve installed at the check valve and fire department connection to keep the valve and connection dry and operating properly during freezing conditions (442)
27. *Sprinklers* — To discharge water after the release of a cap or plug that is activated by some heat-responsive element (434-435)
28. *Control valves* — To cut off the water supply to the system so that sprinklers can be replaced, maintenance performed, or operations interrupted (438)
29. *Alarm test valve* — To simulate actuation of the system by allowing water to flow into the retard chamber and operate the waterflow alarm devices (440)
30. *Inspector's test valve* — To simulate the activation of one sprinkler (440)
31. *Waterflow alarms* — To alert building occupants that water is flowing within the system; some types also notify the fire department (440)
32. *Wet-pipe system*
    A. *Use* — In locations that will not be subjected to freezing temperatures
    B. *Operation* — This system contains water under pressure at all times and is connected to the water supply so that a fused sprinkler will immediately discharge a water spray in that area and actuate an alarm. To shut down the system, the main water control valve should be turned off and the main drain opened. (444)
33. *Dry-pipe system*
    A. *Use* — In locations where piping may be subjected to freezing conditions
    B. Operation — In this system, air under pressure replaces water in the sprinkler piping above the dry-pipe valve; when a sprinkler fuses, the pressurized air escapes first, and then the dry-pipe valve automatically opens to permit water into the piping system. (446)
34. *Pre-action system*
    A. *Use* — In situations where it is especially important that water damage be prevented, even if pipes should be broken
    B. *Operation* — In this dry system that employs a deluge-type valve, fire detection devices, and closed sprinklers, the actuation of the smoke or heat detection system operates a release to open the deluge valve and permits water to enter the distribution system so that water is ready when the sprinklers fuse. (446)

**13**

35. *Deluge system*
    A. *Use* — To protect extra-hazardous occupancies
    B. *Operation* — In this system equipped with open sprinklers and a deluge valve, fire and heat detecting devices or smoke detecting devices trip the device responsible for activating the system. (446)
36. Cross main (435)
37. Riser (435)
38. Fire department connection (435)
39. Main drain connection (435)
40. Water supply (435)
41. Feeder main (435)
42. Alarm valve (435)
43. OS&Y valve to control water supply to system (435)
44. Inspector's test connection (435)
45. Automatic sprinkler heads (435)

## LISTING

46. The following are correct:
    - Partially or completely closed main water control
    - Interruption to the municipal water supply
    - Damaged or painted-over sprinklers
    - Frozen or broken pipes
    - Excess debris or sediment in the pipes
    - Failure of a secondary water supply (434)
47. A. Public water system (440)
    B. Gravity tanks (440)
    C. Pressure tanks (440)
    D. Fire pumps (441)
    E. Fire department connections (441)

## SHORT ANSWER

48. A complete sprinkler system protects the entire building; a partial sprinkler system protects only certain areas such as high hazard areas, exit routes, or places designated by code or by the authority having jurisdiction. (433)
49. The maximum temperature expected at the level of the sprinkler under normal conditions and the anticipated rate of heat release produced by a fire in the particular area (435)
50. By the opening of the sprinklers' release mechanisms, such as fusible links, glass bulbs, chemical pellets, and quick-response mechanisms, in response to the presence of heat (436)
51. Immediately under the sprinkler alarm valve, under the dry-pipe or deluge valve, or outside the building near the sprinkler system that it controls (438)
52. Out of the yoke (439)

**13**

53. To prevent water from flowing from the sprinkler system back into the fire department connection (441)
54. By arrows on the valve or by observing the appearance of the valve casting (442)
55. That the system has been tripped and water has been allowed to enter the pipes (444)
56. That such action could lead to a dramatic increase in the intensity of the fire. A firefighter with a portable radio should be stationed at the control valve in case the valve needs to be reopened. (448-449)
57. By inserting small wooden wedges between the sprinkler discharge orifice and the deflector and tapping them together by hand until the flow is stopped or by inserting commercially made stoppers to plug the orifice (449)

# Chapter 14 Answers

**14**

## DEFINITIONS OF KEY TERMS

1. *Salvage* — Those methods and operating procedures allied to fire fighting that aid in reducing fire, water, and smoke damage during and after fires (453)
2. *Overhaul* — The practice of searching a fire scene to detect hidden fires or sparks which may rekindle and to note the possible point of origin and cause and of recognizing and preserving any evidence of arson (453)

## TRUE/FALSE

3. True (457)
4. False. *Two firefighters are* needed to fold a salvage cover for a one-firefighter spread. (461)
5. True (466)
6. False. By using it as a *chute*, a salvage cover may be used to catch and route water away from fire fighting operations. (467)
7. True (467)
8. True (471)
9. False. Typically, *booster lines or 1½-inch (38 mm)* attack lines are used for overhaul. (472)

## MULTIPLE CHOICE

10. A (453)
11. C (453)
12. B (453)
13. B (458)
14. C (469)
15. A (469)
16. C (469)
17. C (470)
18. D (470)
19. C (473)
20. A (474)

**14**

21. *Sprinkler kit* — A group of tools used to stop the flow of water from open sprinklers (457)

22. *Carryall* — Bag usually made from an old or damaged salvage cover that is used to carry debris, catch falling debris, and provide a water basin for immersing small burning objects (457)

23. *Catchall* — Container constructed from a salvage cover that has been placed on the floor to hold small amounts of water or to temporarily control large amounts of water until it can be routed outside (467)

24. *Canvas salvage covers* — Clean by showering with a hose stream and scrubbing with a broom — scrubbing extremely dirty and stained areas with a detergent solution — and then rinsing thoroughly. Always clean before allowing them to dry and allow to dry before folding. Patch or tape holes. (456)

25. *Synthetic salvage covers* — Clean by rinsing off; it is better to allow them to dry before folding them, but they may be folded wet. Patch or replace if torn. (456)

26. *Water vacuum* — Inspect the power cord for broken insulation, flush the collection tank, and clean the nozzle. (459)

27. *Mops* — Clean with soap and water, and dry thoroughly. (459)

28. *Buckets and tubs* — Clean them and check them for holes. (459)

29. *Tools* — Make sure tools are dry, and lightly oil them if needed. (459)

30. *Brooms* — Clean them and sand the handles if they are burred. (459)

31. *By sound* — Listening for popping or cracking of burning or for hissing of steam (470)

32. *By touch* — Feeling walls and floors with the back of the hand (470)

33. *By sight* — Looking for discoloration of materials, peeling paint, smoke emissions from cracks, cracked plaster, and dried wallpaper (470)

34. *By using electronic sensors* — Using to detect heat variances in objects that may be in hidden locations through thermal imaging (470)

35. A. Guard the evidence where it is found, untouched and undisturbed, to preserve the chain of custody.
    B. Properly identify, remove, and safeguard evidence that cannot be left at the scene of the fire. (472)

36. A. Sprinkler wrench
    B. Sprinkler tong or stopper
    C. Sprinkler wedge (457)

37. The following are correct:
    - Weakened floors due to floor joists being burned away
    - Concrete that has spalled due to the heat
    - Weakened steel roof members
    - Walls offset due to elongation of steel roof supports
    - Weakened roof trusses due to burn-through of key members
    - Mortar in wall joints opened due to excessive heat
    - Wall ties holding veneer walls melted from heat (469-470)

## SHORT ANSWER

38. Near the center of the room in a group or in close piles that can be covered with a minimum of salvage covers (454)
39. Lack of skids or pallets under stock susceptible to water damage, not enough space between stock and the ceiling to allow easy application of salvage covers, shelves that are built to the ceiling and directly against the wall (455)
40. The intensity of the fire and the amount of water used for its control (469)
41. In the area of actual fire involvement and as soon as possible after the fire has been knocked down and the person responsible for determining the cause and origin of the fire says it is all right (470)
42. Between the area being pulled and the doorway to keep the exit route from being blocked with falling debris (471)
43. By pulling off the molding to expose the inner parts (470)
44. Firefighters can submerge an entire small object in a container of water rather than drenching it with a stream of water and can remove a larger item outside to extinguish it. (472)

# Chapter 15 Answers

## DEFINITIONS OF KEY TERMS

1. *Load-bearing wall* — Wall that supports structural weight (480)
2. *Nonload-bearing wall* — Wall that does not support structural weight (480)
3. *Party wall* — A load-bearing wall that supports two adjacent structures (480)
4. *Partition wall* — A nonload-bearing wall that divides two areas within a structure (480)
5. *Fire wall* — Wall that separates two connected structures or that divides portions of a large structure and prevents the spread of fire from one structure, or portion, to the next (481)

## TRUE/FALSE

6. False. It is the obligation of *all personnel, even though a specific safety officer may be designated at the scene,* to constantly monitor for unsafe conditions. (484)
7. False. Wood shake shingles are *not* effective in stopping the spread of fire. (485)
8. False. Firefighters can expect steel structural members to fail at temperatures near or above *1,000°F (540°C)*. (482)
9. True (485)
10. True (486)
11. True (487)

**15**

12. A (485)
13. D (485)
14. B (488)

**IDENTIFICATION**

15. *NFPA 220 — Standard on Types of Building Construction* (477)
16. *Reinforced concrete* — Concrete that is internally fortified with steel reinforcement bars or mesh (483)
17. *Heavy content fire loading* — The storing of combustible materials in high piles that are placed close together (484)
18. *Type I construction* — Fire-resistive construction. This type has structural members that are made of noncombustible or limited combustible materials. Primary fire hazards are the contents of the structure. The ability of this type to confine fire to an area can be compromised by openings made in partitions and improperly designed and dampered central heating and air-conditioning systems. (477)
19. *Type II construction* — Noncombustible or limited combustible construction. This type is similar to Type I except that the degree of fire resistance is lower; materials with no fire-resistance ratings may be used. The primary fire concern is the contents of the building. Buildings of this type often have flat, built-up roofs. (478-479)
20. *Type III construction* — Ordinary construction. This type has exterior walls and structural members that are noncombustible or limited combustible materials and has interior structural members completely or partialy constructed of wood. The primary fire concern is the problem of fire and smoke spread through concealed spaces. (479)
21. *Type IV construction* — Heavy timber construction. This type has exterior and interior walls and their associated structural members of noncombustible or limited combustible materials. Other interior structural members are made of solid or laminated wood with no concealed spaces. The primary concern is the massive amount of combustible contents (fire loading) presented by the structural timbers in addition to the contents of the building. (479)
22. *Type V construction* — Wood-frame construction. This type has exterior walls, bearing walls, floors, roofs, and supports made completely or partially of wood or other approved materials of smaller dimensions than those used for Type IV construction. Construction of this type presents almost unlimited potential for fire extension within the building of origin and to nearby structures. (480)
23. *Wood* — How it is affected by fire depends on the size of the wood (the smaller the wood size, the more likely it is to lose structural integrity) and the moisture content of the wood (that with a high moisture content will not burn as fast as wood that has been cured or dried). The application of water does not substantially affect its structural integrity. (480-481)
24. *Masonry* — It is minimally affected by fire; stones may spall and blocks may crack, but they usually retain most of their strength and stability. The mortar may be subject to more deterioration. Rapid cooling — that occurs when applying water for extinguishment — may cause spalling and cracking. (482)

4-93

25. *Cast iron* — It stands up well during fire. However, the bolts or connections that hold it to the building can fail due to the fire and heat, causing the cast iron sections to fall. (482)

26. *Steel* — Steel elongates when heated. Therefore, it may buckle and fail if its ends are restricted from movement, or it is possible that steel members may push out load-bearing walls and cause a collapse. However, if enough water is applied, it will cool the steel. (482)

27. *Reinforced concrete* — It does not stand up well during fire; it loses strength and spalls. The heating may cause the bond between the concrete and the steel reinforcement to fail. (483)

28. *Gypsum* — It has excellent heat-resistant, fire-retardant properties but will gradually break down under fire conditions. Structural members behind failed gypsum will be subjected to higher temperatures and could fail. (483)

29. *Glass* — Wire-reinforced glass may provide some thermal protection as a separation, but generally glass is not an effective barrier to fire extension. The resins used to bind fiberglass are combustible and can be difficult to extinguish. Applying cold water from a fire stream may cause heated glass to crack. (483)

## LISTING

30. The following are correct:
    - Heavy content fire loading (484)
    - Combustible furnishings and finishes (484)
    - Wood shake shingles (485)
    - Wooden floors and ceilings (485)
    - Large, open spaces (485)
    - Building collapse (485)

31. Answer should include at least five of the following:
    - Cracks or separations in walls, floors, ceilings, and roof structures
    - Evidence of existing structural instability such as the presence of tie rods and stars that hold the walls together
    - Loose bricks, blocks, or stones falling from the buildings
    - Deteriorated mortar between the masonry
    - Walls that appear to be leaning in one direction or another
    - Structural members that appear to be distorted or pulling away from walls
    - Fires beneath floors that support heavy machinery or other extreme weight loads
    - Prolonged fire exposure to the structural members
    - Unusual creaks and cracking noises (486)

# 15

32. Because it provides access problems for fire fighting personnel during manual fire fighting operations and because the fire can quickly override the capabilities of a fire sprinkler system in such situations (484)

33. They should leave the building, establish a collapse zone (equal to the height of the building) around the perimeter of the building, and if it is necessary to operate fire streams from within the collapse zone, use unmanned master streams. (486)

# 16

# Chapter 16 Answers

**DEFINITIONS OF KEY TERMS**

1. *Incendiary device* — Any contrivance designed and used to start a fire (495)
2. *Trailer* — A combustible material used to spread fire from one point to another (495)

**TRUE/FALSE**

3. False. In most jurisdictions, the *fire department chief* has the ultimate legal responsibility for fire cause determination. (491)
4. False. If a fire is of suspicious origin, *each firefighter* should write a chronological account of important circumstances *personally observed (not hearsay or conjecture)*. (496)
5. True (496)
6. False. The fire department has the authority to bar access to any building *during fire fighting and as long afterward as is deemed reasonably necessary*. (497)
7. True (497)

**MULTIPLE CHOICE**

8. A (496)
9. C (496)
10. C (497)

**IDENTIFICATION**

11. *Time of day* — The people and circumstances that should be found at the scene (492)
12. *Weather* — Whether the heating or air-conditioning system should be operating and whether or not it would be likely for the windows to be wide open (492)
13. *Man-made barriers* — An attempt on someone's part to delay fire fighting efforts (493)

14. *Color of smoke* — What is burning (493)
15. *Color of flame* — The intensity of the fire and together with color of smoke, what is burning (493)
16. *Unusual odors* — The presence of an accelerant (494)
17. Protect and preserve evidence at the scene, use the evidence to determine the exact cause, and properly document the entire process. (491)
18. The following are correct:
    - Unusually fast-spreading fire (493)
    - Discarded containers found inside or outside the structure (494)
    - Abnormal behavior of fire when water is applied (494-495)
    - Areas of uneven burning, local heavy charring, or charring in unusual places (495)
    - High heat intensity (495)
19. The following are correct:
    - Locking gates and possibly having them watched by one person
    - Cordoning off the area and marking with signs
    - Piling goods and materials around the entrance to a small business or plant to discourage entry
    - Employing a full-time guard
    - Completely closing all doors, windows, or other entrances with plywood or similar material (497)

## LISTING

20. A. Fuel ignited
    B. Form of heat of ignition
    C. Source of heat of ignition)
    D. The act or omission by a person that helped to bring all the factors together (491)
21. The following are correct:
    - Time of day
    - Weather and natural hazards
    - Man-made barriers
    - People leaving the scene (492-493)

## SHORT ANSWER

22. Because investigators are seldom present while the firefighters fight the fire, perform overhaul, and interview occupants and witnesses for report information (491)
23. That the fire could have been set to conceal another crime (493, 494)
24. As soon as the last firefighter leaves the scene (497)
25. Unless consent is given by the building owner or occupant, fire department personnel must obtain a search warrant to reenter a scene. (497)

# 17 | Chapter 17 Answers

## TRUE/FALSE

1. True (504)
2. True (504)
3. False. *Rate-of-rise detectors* tend to react more quickly than *fixed-temperature detectors*, but they are not quite as reliable. (508)
4. True (509-510)
5. True (515)

## MULTIPLE CHOICE

6. B (502)
7. A (502)
8. B (504)
9. C (505)
10. C (505-506)
11. D (506)
12. C (507)
13. B (510)
14. C (512)

## IDENTIFICATION

15. *Communications center* — A building or portion of a building that houses the equipment and personnel to receive alarms and dispatch resources (501)
16. *Public alerting systems* — Those systems that may be used by anyone to report an emergency (504)
17. *NFPA 1221 — Standard for the Installation, Maintenance and Use of Public Fire Service Communication Systems* (504)
18. *Enhanced 9-1-1 (E-9-1-1) system* — System of notification that combines telephone and computer equipment to provide the dispatcher with instant information on the location from which the call is being made (504)
19. *Private fire alarm signaling systems* — Those systems that are used to detect and transmit alarms to a fire department communications center (506)
20. *FCC* — Federal Communications Commission; the authorizing agency for all radio communication in the United States (512)
21. *Telephone*
    * Dial the appropriate number (fire department 7-digit number, 9-1-1, "O"perator).
    * Give address, with cross street or landmark if possible.
    * State your name and location.
    * Give the telephone number from which you are calling.
    * State the nature of the emergency.
    * Stay on the line if requested to do so by the dispatcher. (503)

22. *Fire alarm pull box*
    - Send signal as directed on the box.
    - Stay at the box until fire personnel arrive so that you can provide the exact location of the emergency to them. (503)
23. *Local alarm box*
    - Send signal as directed on the box.
    - Notify the fire department by telephone using the procedures for reporting by telephone. (503)
24. A. Broadcasting a radio message ordering firefighters to evacuate
    B. Sounding the audible warning devices on the apparatus at the fire scene for an extended period of time (513)

## LISTING

25. The following are correct:
    - Answer calls promptly.
    - Be pleasant and identify the department or company and yourself.
    - Be prepared to take messages.
    - Take accurate messages by including date, time, name of caller, caller's number, message, and your name.
    - Never leave the line open or someone on hold for an extended period of time.
    - Post the message or deliver the message promptly to the person for whom it is intended.
    - Terminate calls courteously; always hang up last. (504)
26. A. Manually activated
    B. Thermal sensitive
    C. Visible products-of-combustion detectors
    D. Invisible products-of-combustion detectors
    E. Flame detectors
    F. Waterflow detectors (506)
27. Answer should include at least five of the following:
    - Avoid unnecessary transmissions. Be brief, accurate, and to the point.
    - Do not transmit until determining if the air is clear.
    - Any unit working at a fire or rescue scene has priority over any other transmission.
    - Do not use profane or obscene language on the air.
    - Hold the microphone 1 to 2 inches (25 mm to 55 mm) from your mouth at a 45-degree angle.
    - Speak calmly, clearly, and distinctly in a natural conversational rhythm at medium speed.
    - Avoid laying the microphone on the seat of the vehicle because the switch may be pressed and cause interference.
    - Do not touch the antenna when transmitting; radio frequency burns might result. (512-513)

**17**

28. The following are correct:
    - Address, particularly if other than the one reported
    - Building and occupancy description
    - Nature and extent of fire
    - Attack mode selected
    - Rescue and exposure problems
    - Instructions to other responding units
    - Location of incident command position (513)
29. The following are correct:
    - Name of the command officer
    - Change in command location
    - Progress (or lack of it) in situation control
    - Direction of fire spread
    - Exposures by direction, height, occupancy, and distance
    - Any problems or needs
    - Anticipated actions — holding, doubtful (513, 515)

## SHORT ANSWER

30. The person transmitting should make the urgency clear to the dispatcher; the dispatcher should give an attention tone, if used in the system, advise all other units to stand by, and then advise the caller to proceed with the emergency traffic. (513)
31. When command personnel decide that all firefighters should be pulled from within a burning building or other hazardous area because conditions have deteriorated beyond the point of reasonable safety (513)

**18**

# Chapter 18 Answers

## DEFINITIONS OF KEY TERMS

1. *Plot plan* — Sketch showing the general arrangement of the property with respect to streets, other buildings, and important features that will help determine fire fighting procedures (525)
2. *Fire hazard* — A condition that will encourage a fire to start or will increase the extent or severity of the fire (526)
3. *Personal hazards* — All individual traits, habits, and personalities of the people who work, live, or visit the property or building in question (528)

## TRUE/FALSE

4. True (521)
5. False. When making an inspection, firefighters should enter the premises at the main entrance *and obtain permission* to make the inspection. (522)
6. True (526)
7. False. Firefighters *may or may not* test the fire protection systems in the occupancies they inspect. (529)
8. True (536)
9. False. Smoke detector batteries should be changed at least *twice a year or more if necessary.* (536)
10. False. A test button on a smoke detector *may check only the device's horn circuit or on some types may check the detector's sensitivity to smoke.* (537)
11. True (538)
12. False. All fire exit drill alarms should be sounded on the *fire alarm system and not on the signal system used to dismiss classes.* (538)

## MULTIPLE CHOICE

13. B (519)
14. C (520)
15. D (520)
16. B (522)
17. C (523)

18. D (527)
19. B (531)
20. C (532-533)
21. C (533)

22. C (534)
23. A (534)
24. C (536)
25. A (536)

## IDENTIFICATION

26. *NFPA 1031 — Standard for Professional Qualifications for Fire Inspector* (520)
27. *NFPA 1035 — Standard for Professional Qualifications for Public Fire Educator* (520)
28. *NFPA 31 — Standard for the Installation of Oil Burning Equipment* (534)
29. *Common fire hazard* — A fire hazard condition that is prevalent in almost all occupancies and will encourage a fire to start (528)
30. *Special fire hazard* — A fire hazard that arises from the processes or operations that are characteristic of the individual occupancy (528)
31. *Public assembly occupancies*
    The following are correct:
    - Large numbers of people present
    - Insufficient, blocked, or locked exits
    - Highly combustible interior finishes
    - Storage of materials in paths of egress (529)

**18**

32. *Manufacturing occupancies*
    The following are correct:
    - Flammable liquids in dip tanks, ovens, and driers, plus those used in mixing, coating, spraying, and degreasing processes
    - High-piled storage of combustible materials
    - Vehicles, such as fork trucks and other trucks, inside the building
    - Large, open areas
    - Large-scale use of flammable and combustible gases (528)

33. *Commercial occupancies*
    The following are correct:
    - Large amounts of contents
    - Mixed variety of contents
    - Difficulties in entering during closed periods
    - Common attics and cocklofts in many multiple occupancies (528)

34. Poor housekeeping; heating, lighting, and power equipment; floor-cleaning compounds; packing materials; fumigation substances; and other flammable and combustible liquids (528)

35. Only the basement, attic, utility room, storage rooms, kitchen, and garage; other rooms may be inspected if requested by householder or in order to access fuse boxes or breaker panels. (532)

## LISTING

36. A. To become familiar with the buildings and their associated hazards
    B. To visualize how existing strategies apply to the building
    C. To recognize hazards that were not noted before
    D. To provide value to citizens as an aid to prevention of fire
    E. To gain valuable on-site information for the pre-incident plan
    F. To enforce fire and building codes (519-520)

37. Answer should include at least six of the following:
    - Location of fire hydrants, fire alarm boxes, and exposures
    - Condition of the streets
    - General housekeeping of the area surrounding the occupancy
    - Type of buildings and occupancies
    - General appearance of the neighborhood
    - Address numbers for visibility
    - All sides of the building for accessibility
    - Forcible entry problems posed by barred windows or high-security doors
    - Overhead obstructions that would restrict aerial ladder operation (522)

**18**

38. The following are correct:
    - Name of business
    - Type of occupancy
    - Date of inspection
    - Name of people conducting the inspection
    - Name of business owner/occupant
    - Name of property owner
    - Edition of applicable code, as reference for future inspections
    - List of violations and their locations stated in specific terms (Code Section numbers should be referenced.)
    - Specific recommendations for correcting each violation
    - Date of the follow-up inspection (526)
39. The following are correct:
    - Fixed extinguishing systems (529)
    - Portable extinguishers (530)
    - Standpipe systems (530)
    - Fire detection and alarm systems (530)
40. The following are correct:
    - Standpipe systems to ensure that they are ready for service
    - Outlets for workable hose threads
    - Outlets for foreign objects that may have been inserted into them
    - Standpipe cabinet to see if hose and nozzles are there and to check their condition
    - Pressure reducing valves (PRVs) to make sure that they are set at a workable level (530)
41. Answer should include at least four of the following:
    - Maintain a courteous attitude on all inspections.
    - Thank the owner or occupant for the invitation into the home.
    - Remember that the primary interest is preventing a fire that could take the lives of the occupants and destroy the home.
    - Make constructive comments regarding the elimination of hazardous conditions.
    - Keep the inspection confidential.
    - Do not gossip.
    - Never make notes from the inspection available to an insurance carrier, repair service organization, sales promotion groups, or any publicity group that would identify a given home. (532)
42. A. Heating appliances
    B. Cooking procedures
    C. Smoking materials
    D. Electrical distribution
    E. Electrical appliances (532)

**18**

43. Answer should include at least five of the following:
    - Have two (or more) escape exits from every room.
    - Windows should be easily opened by anyone and doors should remain closed.
    - Always stay low if awakened by smoke; do not raise up.
    - Have a whistle by every bed to alert other family members if awakened by the smell of smoke.
    - Roll out of bed and crawl to the door. Feel the door; if it is warm, use the window for escape.
    - Never return to the house once outside.
    - Keep a fire escape ladder by the window in a second-story house, and have all family members practice descending it.
    - Agree upon a meeting place outside the home so that all members can be accounted for after escaping.
    - Ask a neighbor to call the fire department. (534-535)

## SHORT ANSWER

44. Pre-incident planning is the whole process of gathering information, developing procedures, and maintaining information resource systems; the building inspection is the fact-gathering part of pre-incident planning. (519)

45. Within a few weeks after the occupant or the owner has received the written report (523-524)

46. When it is necessary to show elevation changes, mezzanines, balconies, or other structural features (525)

47. A report is an accurate account used to describe a specific state or condition; a record is a report that is stored and is capable of being retrieved upon request. (526)

48. The firefighter should check to make sure that all storage is removed from exit aisles, corridors, or passageways; check the operation of all exit doors; and make sure that any chains, deadbolts, or other extra locking devices are removed immediately. (528)

49. Because people have the opportunity to become aware of fire department programs and activities and more familiar with the duties and responsibilities of firefighters and because inspections give people a complete service — not just an emergency service (531)

50. To reduce the number of fire deaths and home fires (531)

51. To STOP immediately, DROP to the ground, and ROLL around until the flames are smothered. They should not run because running will only fan and worsen the flames. (535)

52. They can be installed easily, and they will operate during power failures. (536)

53. At the minimum, in the hallway outside each sleeping area and between the sleeping area and other rooms in the house. They should be mounted on the ceiling or as high on the walls as possible. (537)

54. To ensure orderly exit under controlled supervision (538)

# Chapter 19 Answers

## DEFINITIONS OF KEY TERMS

1. *Accident* — An unplanned, uncontrolled event resulting from unsafe acts and/or unsafe occupational conditions, either of which can result in injury (543)
2. *Injury* — A hurt, damage, or loss sustained as a result of an accident (543)
3. *Law* — Rule of conduct that is adopted and enforced by an authority having jurisdiction (544)
4. *Standards* — Criterion documents that are developed to serve as models or examples of desired performance or behaviors (544)
5. *Goal* — A broad, general, nonmeasurable statement of desired achievement (544)
6. *Objective* — A specific, measurable, achievable statement of intended accomplishment within a specified time frame (544)
7. *Inverter* — Type of power plant; a step-up transformer that converts 12- or 24-volt DC current from a vehicle into 110- or 220-volt AC current (562)
8. *Generator* — Type of power plant that is powered by a small gasoline or diesel engine and generally having 110- and/or 220-volt capacities (562)

## TRUE/FALSE

9. True (544)
10. False. A physical fitness program will improve general health and *will help reduce stress-related injuries.* (546)
11. False. To improve the efficiency of their cardiovascular systems, firefighters should train *two to three* times a week at a level sufficient to raise their pulse rates within a specific range. (547)
12. True (546-547)
13. True (548)
14. False. A person who smokes cigarettes has a significantly greater chance of developing lung cancer than does a nonsmoker, *and* a smoker is *two to three times more likely* to have a heart attack than is a nonsmoker. (550)
15. True (550)
16. True (550)
17. True (555)
18. False. Natural gas service to a building may be halted by turning the petcock on the gas meter so that it is *perpendicular* to the pipe. (556)
19. True (556)
20. True (560)
21. False. With an unenclosed jump seat, *safety gates* are more effective than *safety bars*. (567)
22. True (567)
23. True (568)

## 19

**MULTIPLE CHOICE**

| | | |
|---|---|---|
| 24. A (549) | 29. C (556) | 34. D (559) |
| 25. C (549) | 30. D (557) | 35. D (565) |
| 26. C (551) | 31. B (558) | 36. A (565-566) |
| 27. A (552) | 32. C (557) | 37. B (568) |
| 28. D (554) | 33. A (557-558) | 38. C (569) |

**IDENTIFICATION**

39. *NFPA 1500 — Standard on Fire Department Occupational Safety and Health Program* (544)
40. *NFPA 1021 — Standard for Fire Officer Professional Qualifications* (551)
41. *ICS* — Incident Command System; system adopted by the National Fire Academy consisting of procedures for controlling personnel, facilities, equipment, and communications at emergency incidents (551-552)
42. *Support branch* — One of two branches of the ICS logistics functional area that includes medical, communications, and food service (552-553)
43. *Service branch* — One of two branches of the ICS logistics functional area that includes supplies, facilities, and ground support (vehicle services) (553)
44. *GFCI* — Ground fault circuit interrupter; device that quickly shuts off electricity in order to prevent or reduce the severity of shock to the user if a tool's insulation fails (560)
45. *Command* — All incident activities, including the development and implementation of strategic decisions (552)
46. *Operations* — Managing all operations directly applicable to the primary mission (552)
47. *Planning* — Collecting, evaluating, disseminating, and using information concerning the development of the incident and maintaining the status of resources (552)
48. *Logistics* — Providing the facilities, services, and materials necessary to support the incident (552)
49. *Finance* — All costs and financial aspects of the incident (554)
50. The following are correct:
    - Try to remain calm.
    - If equipped with a radio, try to make radio contact as quickly as possible with others on the emergency scene.
    - Try to retrace steps or try to seek an exit from the building or at least from the area that is on fire.
    - Shout for help every once in a while.
    - Activate PASS device.
    - Be alert for a hoseline or safety line, and follow it — the male coupling signifies the exit direction. (555)

**LISTING**

51. Answer should include at least five of the following:
    - Wear appropriate personal protective equipment.
    - Remove jewelry, including rings and watches.
    - Select the appropriate tool for the job.
    - Know the manufacturer's instructions and follow them.
    - Inspect tools before use to determine their condition. If a tool has deteriorated or is broken, replace it.
    - Provide adequate storage space for tools, and always return them promptly to storage after use.
    - Consult with and secure the approval of the manufacturer before modifying the tool.
    - Use spark-resistant tools when working in flammable atmospheres such as around a vehicle's fuel system. (559-560)

52. The following are correct:
    - Match the saw to the task and the material to be cut. Never push a saw beyond its design limitations.
    - Wear proper protective equipment, including gloves and eye protection.
    - Have hoselines in place when forcing entry into an area where fire is suspected, when performing vertical ventilation, or when cutting materials that generate sparks.
    - Avoid the use of all saws when working in a flammable atmosphere or near flammable liquids.
    - Keep unprotected and unessential people out of the work area.
    - Follow manufacturer's guidelines for proper saw operation.
    - Use caution to avoid igniting gasoline vapors when refueling a hot, gasoline-powered saw.
    - Keep blades and chains well sharpened. (561)

**SHORT ANSWER**

53. One copy should be kept by the company officer, and one copy should be kept in the apparatus. (554)

54. Firefighters are equipped with personal identification tags which they leave at a given location or with a designated person when they enter the fireground perimeter. These tags are kept on a control board or personnel ID chart. Upon leaving the fireground perimeter, the firefighters collect their tags. (554)

55. An SCBA tag system provides closer accountability for personnel inside the structure. Each firefighter entering a hazardous atmosphere gives his or her tag, which includes user's name and air pressure, to a designated supervisor, who records entry time and expected exit time. Relief crews can then be sent in before the expected exit time of the crew already within the structure. (554-555)

56. When a life is in immediate danger, a rescue must be performed, and the rescuer has the proper knowledge and equipment (557)

**19**

57. Because they operate at a very high temperature and even brief contact with a person can result in serious burns (563)

58. Because overtaxing the power plant will give poor lighting, may damage the power generating unit or the lights, and will restrict the operation of other electrical tools (564)

59. Slip-and-fall accidents and accidents resulting from improper lifting techniques (568)

# IFSTA MANUALS AND FPP PRODUCTS

**For a current catalog describing these and other products call or write your local IFSTA distributor or Fire Protection Publications, IFSTA Headquarters, Oklahoma State University, Stillwater, OK 74078-0118.**
**Phone: 1-800-654-4055**

## FIRE DEPARTMENT AERIAL APPARATUS
includes information on the driver/operator's qualifications; vehicle operation; types of aerial apparatus; positioning, stabilizing, and operating aerial devices; tactics for aerial devices; and maintaining, testing, and purchasing aerial apparatus. Detailed appendices describe specific manufacturers' aerial devices. 1st Edition (1991), 416 pages, addresses NFPA 1002.

## STUDY GUIDE FOR AERIAL APPARATUS
The companion study guide in question and answer format. 1st Edition (1991), 152 pages.

## AIRCRAFT RESCUE AND FIRE FIGHTING
comprehensively covers commercial, military, and general aviation. All the information you need is in one place. Subjects covered include: personal protective equipment, apparatus and equipment, extinguishing agents, engines and systems, fire fighting procedures, hazardous materials, and fire prevention. Over 240 photographs and two-color illustrations. It also contains a glossary and review questions with answers. 3rd Edition (1992), 272 pages, addresses NFPA 1003.

## BUILDING CONSTRUCTION RELATED TO THE FIRE SERVICE
helps firefighters become aware of the many construction designs and features of buildings found in a typical first alarm district and how these designs serve or hinder the suppression effort. Subjects include construction principles, assemblies and their resistance to fire, building services, door and window assemblies, and special types of structures. 1st Edition (1986), 166 pages, addresses NFPA 1001 and NFPA 1031, levels I & II.

## CHIEF OFFICER
lists, explains, and illustrates the skills necessary to plan and maintain an efficient and cost-effective fire department. The combination of an ever-increasing fire problem, spiraling personnel and equipment costs, and the development of new technologies and methods for decision making requires far more than expertise in fire suppression. Today's chief officer must possess the ability to plan and administrate as well as have political expertise. 1st Edition (1985), 211 pages, addresses NFPA 1021, level VI.

## SELF-INSTRUCTION FOR CHIEF OFFICER
The companion study guide in question and answer format. 1st Edition, 142 pages.

## FIRE DEPARTMENT COMPANY OFFICER
focuses on the basic principles of fire department organization, working relationships, and personnel management. For the firefighter aspiring to become a company officer, or a company officer wishing to improve management skills, this manual helps develop and improve the necessary traits to effectively manage the fire company. 2nd Edition (1990), 278 pages, addresses NFPA 1021, levels I, II, & III.

## COMPANY OFFICER STUDY GUIDE
The companion study guide in question and answer format. Includes problem applications and case studies. 1st Edition (1991), 256 pages.

## ESSENTIALS OF FIRE FIGHTING
is the "bible" on basic firefighter skills and is used throughout the world. The easy-to-read format is enhanced by 1,500 photographs and illustrations. Step-by-step instructions are provided for many fire fighting tasks. Topics covered include: personal protective equipment, building construction, firefighter safety, fire behavior, portable extinguishers, SCBA, ropes and knots, rescue, forcible entry, ventilation, communications, water supplies, fire streams, hose, fire cause determination, public fire education and prevention, fire suppression techniques, ladders, salvage and overhaul, and automatic sprinkler systems. 3rd Edition (1992), addresses NFPA 1001.

## STUDY GUIDE FOR 3rd EDITION OF ESSENTIALS OF FIRE FIGHTING
The companion learning tool for the new 3rd edition of the manual. It contains questions and answers to help you learn the important information in the book. 1st Edition (1992).

## PRINCIPLES OF EXTRICATION
leads you step-by-step through the procedures for disentangling victims from cars, buses, trains, farm equipment, and industrial situations. Fully illustrated with color diagrams and more than 500 photographs. It includes rescue company organization, protective clothing, and evaluating resources. Review questions with answers at the end of each chapter. 1st Edition (1990), 400 pages.

## FIRE CAUSE DETERMINATION
gives you the information necessary to make on-scene fire cause determinations. You will know when to call for a trained investigator, and you will be able to help the investigator. It includes a profile of firesetters, finding origin and cause, documenting evidence, interviewing witnesses, and courtroom demeanor. 1st Edition (1982), 159 pages, addresses NFPA 1021, Fire Officer I, and NFPA 1031, levels I & II.

## FIRE SERVICE FIRST RESPONDER
provides the information needed to evaluate and treat patients with serious injuries or illnesses. It familiarizes the reader with a wide variety of medical equipment and supplies. **First Responder** applies to safety, security, fire brigade, and law enforcement personnel, as well as fire service personnel, who are required to administer emergency medical care. 1st Edition (1987), 340 pages, addresses NFPA 1001, levels I & II, and DOT First Responder.

## FORCIBLE ENTRY
reflects the growing concern for the reduction of property damage as well as firefighter safety. This comprehensive manual contains technical information about forcible entry tactics, tools, and methods, as well as door, window, and wall construction. Tactics discuss the degree of danger to the structure and leaving the building secure after entry. Includes a section on locks and through-the-lock entry. Review questions and answers at the end of each chapter. 7th Edition (1987), 270 pages, helpful for NFPA 1001.

## GROUND COVER FIRE FIGHTING PRACTICES

explains the dramatic difference between structural fire fighting and wildland fire fighting. Ground cover fires include fires in weeds, grass, field crops, and brush. It discusses the apparatus, equipment, and extinguishing agents used to combat wildland fires. Outdoor fire behavior and how fuels, weather, and topography affect fire spread are explained. The text also covers personnel safety, management, and suppression methods. It contains a glossary, sample fire operation plan, fire control organization system, fire origin and cause determination, and water expansion pump systems. 2nd Edition (1982), 152 pages.

## FIRE SERVICE GROUND LADDER PRACTICES

is a "how to" manual for learning how to handle, raise, and climb ground ladders; it also details maintenance and service testing. Basic information is presented with a variety of methods that allow the readers to select the best method for their locale. The chapter on Special Uses includes: ladders as a stretcher, a slide, a float drag, a water chute, and more. The manual contains a glossary, review questions and answers, and a sample testing and repair form. 8th Edition (1984), 388 pages, addresses NFPA 1001.

## HAZARDOUS MATERIALS FOR FIRST RESPONDERS

provides basic information on hazardous materials with sections on site management and decontamination. It includes a description of various types of materials, their characteristics, and containers. The manual covers the effects of weather, topography, and environment on the behavior of hazardous materials and control efforts. Pre-incident planning and post-incident analysis are covered. 1st Edition (1988), 357 pages, addresses NFPA 472, 29 CFR 1910.120 and NFPA 1001.

## STUDY GUIDE FOR HAZARDOUS MATERIALS FOR FIRST RESPONDERS

The companion study guide in question and answer format also includes case studies that simulate incidents. 1st Edition (1989), 208 pages.

## HAZARDOUS MATERIALS: MANAGING THE INCIDENT

takes you beyond the basic information found in **Hazardous Materials for First Responders**. Directed to the leader/commander, this manual sets forth basic practices clearly and comprehensively. Charts, tables, and checklists guide you through the organization and planning stages to decontamination. This text, along with the accompanying workbook and instructor's guide, provides a comprehensive learning package. 1st Edition (1988), 206 pages, helpful for NFPA 1021.

## STUDENT WORKBOOK FOR HAZARDOUS MATERIALS: MANAGING THE INCIDENT

provides questions and answers to enhance the student's comprehension and retention. 1st Edition (1988), 176 pages.

## INSTRUCTOR'S GUIDE FOR HAZARDOUS MATERIALS: MANAGING THE INCIDENT

provides lessons based on each chapter, adult learning tips, and appendices of references and suggested audio visuals. 1st Edition (1988), 142 pages.

## HAZ MAT RESPONSE TEAM LEAK AND SPILL GUIDE

contains articles by Michael Hildebrand reprinted from *Speaking of Fire's* popular Hazardous Materials Nuts and Bolts series. Two additional articles from *Speaking of Fire* and the hazardous material incident SOP from the Chicago Fire Department are also included. 1st Edition (1984), 57 pages.

## EMERGENCY OPERATIONS IN HIGH-RACK STORAGE

is a concise summary of emergency operations in the high-rack storage area of a warehouse. It explains how to develop a pre-emergency plan, what equipment will be necessary to implement the plan, type and amount of training personnel will need to handle an emergency, and interfacing with various agencies. Includes consideration questions, points not to be overlooked, and trial scenarios. 1st Edition (1981), 97 pages.

## HOSE PRACTICES

reflects the latest information on modern fire hose and couplings. It is the most comprehensive single source about hose and its use. The manual details basic methods of handling hose, including large diameter hose. It is fully illustrated with photographs showing loads, evolutions, and techniques. This complete and practical book explains the national standards for hose and couplings. 7th Edition (1988), 245 pages, addresses NFPA 1001.

## FIRE PROTECTION HYDRAULICS AND WATER SUPPLY ANALYSIS

covers the quantity and pressure of water needed to provide adequate fire protection, the ability of existing water supply systems to provide fire protection, the adequacy of a water supply for a sprinkler system, and alternatives for deficient water supply systems. 1st Edition (1990), 340 pages.

## INCIDENT COMMAND SYSTEM (ICS)

was developed by a multiagency task force. Using this system, fire, police, and other government groups can operate together effectively under a single command. The system is modular and can be used to meet the requirements of both day-to-day and large-incident operations. It is the approved basic command system taught at the National Fire Academy. 1st Edition (1983), 220 pages, helpful for NFPA 1021.

## INDUSTRIAL FIRE PROTECTION

is designed for the person charged with the responsibility of developing, implementing, and coordinating fire protection. A "must read" for fire service personnel who will coordinate with industry/business for pre-incident planning. The text includes guidelines for establishing a company policy, organization and planning for the emergency, establishing a fire prevention plan, incipient fire fighting tactics, an overview of interior structural fire fighting, and fixed fire fighting systems. 1st Edition (1982), 207 pages, written for 29 CFR. 1910, Subpart L, and helpful for NFPA 1021 and NFPA 1031.

## FIRE INSPECTION AND CODE ENFORCEMENT

provides a comprehensive, state-of-the-art reference and training manual for both uniformed and civilian inspectors. It is a comprehensive guide to the principles and techniques of inspection. Text includes information on how fire travels, electrical hazards, and fire resistance requirements. It covers storage, handling, and use of hazardous materials; fire protection systems; and building construction for fire and life safety. 5th Edition (1987), 316 pages, addresses NFPA 1001 and NFPA 1031, levels I & II.

## STUDY GUIDE FOR FIRE INSPECTION AND CODE ENFORCEMENT

The companion study guide in question and answer format with case studies. 1st Edition (1989), 272 pages.

## FIRE SERVICE INSTRUCTOR

explains the characteristics of a good instructor, shows you how to determine training requirements, and teach to the level of your class. It discusses the types, principles, and procedures of teaching and learning, and covers the use of effective training aids and devices. The purpose and principles of testing as well as test construction are covered. Included are chapters on safety, legal considerations, and computers. 5th Edition (1990), 325 pages, addresses NFPA 1041, levels I & II.

## LEADERSHIP IN THE FIRE SERVICE

was created from the series of lectures given by Robert F. Hamm to assist in leadership development. It provides the foundation for getting along with others, explains how to gain the confidence of your personnel, and covers what is expected of an officer. Included is information on supervision, evaluations, delegating, and teaching. Some of the topics include: the successful leader today, a look into the past may reveal the future, and self-analysis for officers. 1st Edition (1967), 132 pages.

## FIRE SERVICE ORIENTATION AND INDOCTRINATION

relates the traditions, history, and organization of the fire service. It includes operation of the fire department, responsibilities and duties of firefighters, and the function of fire department companies. This exciting and informative text is for anyone dealing with the fire service who needs a basic understanding and overview. The perfect book for new or prospective members, buffs, your congressman or council members, fire service sales personnel, and industrial brigades. 2nd Edition (1984), 187 pages, addresses NFPA 1001.

## PRIVATE FIRE PROTECTION AND DETECTION

introduces the firefighter, inspection personnel, brigade/safety member, insurance inspector/investigator, or fire protection student to fixed systems, extinguishers, and detection. It covers the various types of equipment, their installation, maintenance, and testing. Systems discussed are: wet-pipe, dry-pipe, pre-action, deluge, residential, carbon dioxide, Halogen, dry- and wet-chemical, foam, standpipe, and fire extinguishers. 1st Edition (1979), 170 pages, addresses NFPA 1001 and NFPA 1031, levels I & II.

## PUBLIC FIRE EDUCATION

provides valuable information for ending public apathy and ignorance about fire. This manual gives you the knowledge to plan and implement fire prevention campaigns. It shows you how to tailor the individual programs to your audience as well as the time of year or specific problems. It includes working with the media, resource exchange, and smoke detectors. 1st Edition (1979), 169 pages, helpful for NFPA 1021 and 1031.

## FIRE DEPARTMENT PUMPING APPARATUS

is the Driver/Operator's encyclopedia on operating fire pumps and pumping apparatus. It covers pumpers, tankers (tenders), brush apparatus, and aerials with pumps. This comprehensive volume explains safe driving techniques, getting maximum efficiency from the pump, and basic water supply. It includes specification writing, apparatus testing, and extensive appendices of pump manufacturers. 7th Edition (1989), 374 pages, addresses NFPA 1002.

## STUDY GUIDE FOR PUMPING APPARATUS

The companion study guide in question and answer format. 1st Edition (1990), 100 pages.

## FIRE SERVICE RESCUE PRACTICES

is a comprehensive training text for firefighters and fire brigade members that expands proficiency in moving and removing victims from hazardous situations. This extensively illustrated manual includes rescuer safety, effects of rescue work on victims, rescue from hazardous atmospheres, trenching, and outdoor searches. 5th Edition (1981), 262 pages, addresses NFPA 1001.

## RESIDENTIAL SPRINKLERS A PRIMER

outlines United States residential fire experience, system components, engineering requirements, and issues concerning automatic and fixed residential sprinkler systems. Written by Gary Courtney and Scott Kerwood and reprinted from *Speaking of Fire.*

An excellent reference source for any fire service library and an excellent supplement to **Private Fire Protection.** 1st Edition (1986), 16 pages.

## FIRE DEPARTMENT OCCUPATIONAL SAFETY

addresses the basic responsibilities and qualifications for a safety officer and the minimum requirements and procedures for a safety and health program. Included in this manual is an overview of establishing and implementing a safety program, physical fitness and health considerations, safety in training, fire station safety, tool and equipment safety and maintenance, personal protective equipment, en route hazards and response, emergency scene safety, and special hazards. 2nd Edition (1991), 396 pages, addresses NFPA 1500, 1501.

## SALVAGE AND OVERHAUL

covers planning salvage operations, equipment selection and care, as well as describing methods and techniques for using salvage equipment to minimize fire damage caused by water, smoke, heat, and debris. The overhaul section includes methods for finding hidden fire, protection of fire cause evidence, safety during overhaul operations, and restoration of property and fire protection systems after a fire. 7th Edition (1985), 225 pages, addresses NFPA 1001.

## SELF-CONTAINED BREATHING APPARATUS

contains all the basics of SCBA use, care, testing, and operation. Special attention is given to safety and training. The chapter on Emergency Conditions Breathing has been completely revised to incorporate safer emergency methods that can be used with newer models of SCBA. Also included are appendices describing regulatory agencies and donning and doffing procedures for nine types of SCBA. The manual has been thoroughly updated to cover NFPA, OSHA, ANSI, and NIOSH regulations and standards as they pertain to SCBA. 2nd Edition (1991), 360 pages, addresses NFPA 1001.

## STUDY GUIDE FOR SELF-CONTAINED BREATHING APPARATUS

The companion study guide in question and answer format. 1st Edition (1991).

## FIRE STREAM PRACTICES

brings you an all new approach to calculating friction loss. This carefully written text covers the physics of fire and water; the characteristics, requirements, and principles of good streams; and fire fighting foams. **Streams** includes formulas for the application of fire fighting hydraulics, as well as actions and reactions created by applying streams under a variety of circumstances. The friction loss equations and answers are included, and review questions are located at the end of each chapter. 7th Edition (1989), 464 pages, addresses NFPA 1001 and NFPA 1002.

## GASOLINE TANK TRUCK EMERGENCIES

provides emergency response personnel with background information, general procedures, and response guidelines to be followed when responding to and operating at incidents involving MC-306/DOT 406 cargo tank trucks. Specific topics include: incident management procedures, site safety considerations, methods of product transfer, and vehicle uprighting considerations. 1st Edition (1992), 60 pages, addresses NFPA 472.

## FIRE VENTILATION PRACTICES

presents the principles, practices, objectives, and advantages of ventilation. It includes the factors and phases of combustion, flammable liquid characteristics, products of combustion, backdrafts, transmission of heat, and building construction considerations. The manual reflects the new techniques in building construction and their effects on ventilation procedures. Methods and procedures are thoroughly explained with numerous photographs and drawings. The text also includes: vertical (top), horizontal (cross), and forced ventilation; and a glossary. 6th Edition (1980), 131 pages, addresses NFPA 1001.

## FIRE SERVICE PRACTICES FOR VOLUNTEER AND SMALL COMMUNITY FIRE DEPARTMENTS

presents those training practices that are most consistent with the activities of smaller fire departments. Consideration is given to the limitations of small community fire department resources. Techniques for performing basic skills are explained, accompanied by detailed illustrations and photographs. 6th Edition (1984), 311 pages.

## WATER SUPPLIES FOR FIRE PROTECTION

acquaints you with the principles, requirements, and standards used to provide water for fire fighting. Rural water supplies as well as fixed systems are discussed. Abundant photographs, illustrations, tables, and diagrams make this the most complete text available. It includes requirements for size and carrying capacity of mains, hydrant specifications, maintenance procedures conducted by the fire department, and relevant maps and record keeping procedures. Review questions at the end of each chapter. 4th Edition (1988), 268 pages, addresses NFPA 1001, NFPA 1002, and NFPA 1031, levels I & II.

## TEACHING PACKAGES

### LEADERSHIP

This teaching package is designed to assist the instructor in teaching leadership and motivational skills. Cause and effect, behavior and consequences, listening and communications are themes throughout the course that stress the reality of the job and the people one deals with daily. Before each lesson is a title page that gives an outline of the subject matter to be covered, the approximate time required to teach the material, the specific learning objectives, and the references for the instructor's preparation. Sources for suggested films and videotapes are included.

### CURRICULUM PACKAGE FOR IFSTA COMPANY OFFICER

A competency-based Teaching Package with lesson plans and activities to teach the student the information and skills needed to qualify for the position of Company Officer. Corresponds to **Fire Department Company Officer**, 2nd Edition.

The Package includes the Company Officer Instructor's Guide (the how, what, and when to teach); the Student Guide (a workbook for group instruction or self-study); and 143 full-color overhead transparencies.

### ESSENTIALS CURRICULUM PACKAGE

A competency-based teaching package with lesson plans and activities to teach the student the information and skills needed to qualify for the position of Fire Fighter I or II. Corresponds to **Essentials of Fire Fighting**, 3rd Edition.

The Package, with Instructor's Guide, Student Guide, and more than 400 transparencies, is scheduled for publication in 1993.

## TRANSLATIONS

### LO ESENCIAL EN EL COMBATE DE INCENDIOS

is a direct translation of **Essentials of Fire Fighting**, 2nd edition. Please contact your distributor or FPP for shipping charges to addresses outside U.S. and Canada. 444 pages.

### PRACTICAS Y TEORIA PARA BOMBEROS

is a direct translation of **Fire Service Practices for Volunteer and Small Community Fire Departments**, 6th edition. Please contact your distributor or FPP for shipping charges to addresses outside U.S. and Canada. 347 pages.

## OTHER ITEMS

### TRAINING AIDS

Fire Protection Publications carries a complete line of videos, overhead transparencies, and slides. Call for a current catalog.

### NEWSLETTER

The nationally acclaimed and award winning newsletter, *Speaking of Fire*, is published quarterly and available to you free. Call today for your free subscription.

Study Guide for
**ESSENTIALS of Fire Fighting**
1st EDITION
2nd PRINTING, 4/93

**COMMENT SHEET**

DATE _____ NAME _____

ADDRESS _____

ORGANIZATION REPRESENTED _____

CHAPTER TITLE _____ NUMBER _____

SECTION/PARAGRAPH/FIGURE _____ PAGE _____

1. Proposal (include proposed wording, or identification of wording to be deleted),
   OR PROPOSED FIGURE:

2. Statement of Problem and Substantiation for Proposal:

RETURN TO: IFSTA Editor
Fire Protection Publications
Oklahoma State University
Stillwater, OK 74078

SIGNATURE _____

Use this sheet to make any suggestions, recommendations, or comments. We need your input to make the manuals as up to date as possible. Your help is appreciated. Use additional pages if necessary.